Martianus Capella and the Seven Liberal Arts

VOLUME I

NUMBER LXXXIV OF THE

Records of Civilization: Sources and Studies

mille reliquisq; ſoliditas. Finis ergo vel limites mihi ſunt monas, decas, hecatontas, & mille: Geometriæ verò nota, linea, figura, ſoliditas. Nam monas ita indiuidua eſt, vt nota: Decas verò in numeris, vt linea longitudinis ſolius. Hecatontas quadratus, qui eſt ſuperficies, & in latitudinem diuiditur. Item decas per decas fit centum quadratus. Hoc per decem fit cubus mille. Omnis impar progreſſus à monade per ſingulas poſitiones neceſſariò quadratos efficit. Prima ipſa monas adde triadem fecit quatuor, primũ quadratum: aſſocias quinque, feciſti ſecundum quadratum nouem: iunge 7, impleſti quadratum ſedecim: adijcis item nonas, & perficis quadratum 25. Eodem modo progreditur ratio vſq; in infinitum. Sed ad ſuperius diuiſa regrediar. Omnem numerum aut parem, aut imparem eſſe, aut vtroque finiri. Quicquid numero adijciatur, finito finitum adijci: neque ex finitis finitum fieri poſſe.

De pari & impari, ex iiſque compoſitis.

OMNIS verò numerus, aut par aut impar eſt. Par eſt, qui in duas æquas partes diuiditur, vt 2.4.6. Impar, qui in duas æquas partes diuidi non poteſt, vt 3. 5. 7. Deinde ex imparibus, quidam ex imparibus tantùm impares ſunt, vt 3. 5. 7. Quidam etiam multitudine conſtant, vt 9. 15. 21. quos περισσάκις περισσοὺς Græci appellant. At in his qui pares ſunt, plura diſcrimina ſunt, at pares ſunt & diuidi poſſunt. Cæteri vel ex paribus pares, vel ex paribus impares, vel ex imparibus pares. Et illos Græci ἀρτιάκις ἀρτίους, vel περισσάκις ἀρτίους, vel ἀρτιάκις περισσοὺς nominant. Pares ex paribus ſunt quatuor, quia ex bis binis: octo, quia ex bis quaternis cõſtant. Pares ex imparibus ſunt qui pares impari multiplicatione fiunt, vt bis terni * ſex, aut quinquies quaterni * viginti, quod genus Græci περισσάκις ἀρτίους vocant. Et hi qui imparem numerorum multitudinem pari multiplicatione conſummãt, vt cùm bis ter in ſex, & qua-

Courtesy of the British Museum

PAGE FROM THE GROTIUS EDITION OF MARTIANUS (1599), WITH EMENDATIONS BY RICHARD BENTLEY IN THE MARGIN

Martianus Capella and the Seven Liberal Arts

VOLUME I

THE QUADRIVIUM OF MARTIANUS CAPELLA

LATIN TRADITIONS IN THE MATHEMATICAL SCIENCES

50 B.C.-A.D. 1250

by William Harris Stahl

WITH A STUDY OF THE ALLEGORY

AND THE VERBAL DISCIPLINES

by Richard Johnson with E. L. Burge

COLUMBIA UNIVERSITY PRESS

NEW YORK AND LONDON 1971

The late WILLIAM HARRIS STAHL, editor and translator of *Macrobius' Commentary on the Dream of Scipio*, was Professor of Classics at Brooklyn College.
RICHARD JOHNSON is Professor of Classics at the Australian National University, at Canberra.
E. L. BURGE is a Senior Lecturer in Classics at the Australian National University.

Library of Congress Catalog Card Number: 76-121876
Standard Book Number: 231-03254-4
Printed in the Netherlands

Records of Civilization: Sources and Studies

EDITED UNDER THE AUSPICES OF THE DEPARTMENT OF HISTORY
COLUMBIA UNIVERSITY

Foreword

IT WAS the suggestion of the late William H. Stahl that he and I should join in preparing a translation of Martianus Capella for the Records of Civilization. With the two collaborators ten thousand miles apart, the work extended over almost a decade, Mr. Stahl completed his section of the translation at least two years before I completed mine, and his introduction grew into an independent volume, the first large-scale study of Martianus ever made. Finally I completed my share of the translation and a very modest section of the introduction, having enlisted meantime the scholarship of my colleague E. L. Burge to assist with the difficulties of Book IV.

In December 1968 our separate contributions were all substantially complete. They reflected different approaches and points of emphasis, inevitable in an enterprise undertaken by widely separated collaborators. Before any major revision and coordination could be undertaken, William Stahl died unexpectedly. The publishers have felt (I believe rightly) that it would be improper to make any changes of substance or approach in his writing. Therefore the study as published reflects to some extent the separate hands at work.

It is a matter of great regret to me that, because of delays for which I was responsible, William Stahl never saw published the book which is the capstone of his studies in ancient science.

R. JOHNSON

Canberra, A. C. T.
April 1, 1970

Preface

MARTIANUS CAPELLA is as difficult an author as he is an important one. Each time one reads large portions of his text he gains new insights and is inclined to revise former opinions. Because of the difficulty of the problems that confront investigators and clamor for an answer, little Martianus scholarship of a comprehensive scope has appeared in print. Most earlier studies of the *De nuptiis Philologiae et Mercurii* have dealt with particular books or isolated aspects. Literary critics have examined Books I and II, historians of rhetoric Book V, historians of cosmography Books VI and VIII, historians of mathematics Books VI and VII, and musicologists Book IX. One cannot speak with assurance about Martianus, however, until he has become thoroughly familiar with the whole of the *De nuptiis*—an arduous undertaking indeed. Answers to its problems have to be sought in the work as a whole. The trivium (Books III-V) and quadrivium (Books VI-IX) portions are interdependent, as those disciplines were in the ancient and medieval curriculum; and the description of the setting of the heavenly wedding in Book I and II anticipates many matters contained in the disciplines. The work has rarely been read in its entirety in modern times, except by its editors.

A key figure in the intellectual history of Western Europe, Martianus is deserving of closer examination. Half classical, half medieval, his work may be likened to the neck of an hourglass through which the classical liberal arts trickled to the medieval world. The materials that Martianus handled were the common property of Latin compilers of his age. Citations of parallel passages from this vast store, unless they present verbal correspondences, are of little value as traces of borrowing and are impressive only to readers who are unfamiliar with the habits of compilers. I have tried to restrict my use of parallels to those that illuminate the text at hand or represent a possible line of transmission, as in the case of Calcidius, Macrobius, Boethius, Cassiodorus, and Isidore. Martianus and these five authors were the mine from which medieval Latin science was drawn. Pliny, to be sure, was an equally important figure and had an equally important influence; but it is very difficult to trace the diffusion of the material in his *Natural History*. His bulky work was digested early and it entered the scientific traditions mostly through intermediaries or in excerpts.

I wish here to express my appreciation to the foundations and institutions that provided me with the opportunity and the facilities to study abroad during my sabbatical year 1962-1963—to the John Simon Guggenheim Memorial Foundation for a fellowship; to the Fulbright authorities for a travel grant to Australia; to the Australian National University for a fellowship at University House, Canberra—and to the scholars who greatly assisted me during my sabbatical studies: Professor E. G. Turner of the University of London; Mr. J. B. Trapp and Dr. C. R. Ligota of the Warburg Institute, University of London; and Professors Arthur Dale Trendall and Richard Johnson of the Australian National University. To the last-named, my collaborator on a forthcoming translation and commentary of Martianus' work (Vol. II of this set), I am particularly indebted for his gracious assistance in dealing with problems of interpretation and for his contribution of the part on the trivium to this volume. Mr. Evan Burge of the Australian National University ably assisted him in the preparation of this section. I also wish to thank other tillers of the stony "Martian" soil, Claudio Leonardi, Cora E. Lutz, Jean Préaux, and James Willis, for sending me offprints of their recent studies; Professors Siegmund Levarie and Carl Boyer of Brooklyn College for their abundant and expert guidance in coping with the enigmas of Martianus' exposition of harmony in Book IX and arithmetic and geometry in Books VI and VII; and Dr. Karl-August Wirth for his kind help in correcting and supplementing my bibliography on the medieval and Renaissance iconography of the liberal arts. Lastly, I am grateful to Professor W. T. H. Jackson, Editor of the Records of Civilization, for his many helpful suggestions; to Jacques Barzun, former editor of the series, who in 1959 sponsored the wedding of Philology and Mercury; and to Mrs. Michelle Kamhi of Columbia University Press for the extraordinary skill and care she has brought to bear in editing this manuscript for the press.

W. H. STAHL

Baldwin, N. Y.
March 15, 1969

Contents

Tables

COMPILED FROM BOOK IX OF MARTIANUS

Abbreviations

USED IN THE FOOTNOTES

BISIAM	*Bullettino dell' Istituto storico italiano per il medio evo e Archivio muratoriano*
CMH	*The Cambridge Medieval History*
CP	*Classical Philology*
DHGE	*Dictionnaire d'histoire et de géographie ecclésiastique*
IMU	*Italia medioevale e umanistica*
JWCI	*Journal of the Warburg and Courtauld Institutes*
OCD	*Oxford Classical Dictionary*
Pauly-Wissowa	*Paulys Real Encyclopädie der classischen Altertumswissenschaft*
PL	*Patrologiae cursus completus, series latina*
SIFC	*Studi italiani di filologia classica*
SM	*Studi medievali*
TLL	*Thesaurus Linguae Latinae*

PART I

Introduction

Roman Science in Intellectual History

TO SAY that the Romans had no science is untrue. To adopt the opposite attitude and attempt to represent them as respectable scientists by pointing to their achievements in engineering and technology is equally misleading. The technicians, mechanics, and craftsmen of the ancient world, except in rare instances, did not leave explicit records of the development of their technical knowledge. They had a job to perform and were not interested in the theoretical literature on the subject; recipe books sufficed for their needs. It has been supposed that the engineers in charge of the construction of Roman public and military works were familiar with the mechanical and mathematical treatises of Heron of Alexandria. We may assume, however, that any supervisor or technical consultant who was able to comprehend Heron's original writings had been educated in Greek schools. Latin schools did not produce students who could develop geometrical proofs. The genuine treatises in Heron's collection contain a considerable amount of theory. The numerical and geometrical formulas that Roman engineers used had actually been edited and reedited for practical application by generations of surveyors and architects, although they were still ascribed during the Empire to Heron.[1] Engineering and technology do not provide clues to the competence of Roman scientists.

On the other hand, to examine the extant Latin scientific texts, early and late, and to regard them in their *Vorleben* and *Nachleben*—from

[1] On the rule-of-thumb character of Roman applied science and on Roman dependence upon the Greeks as technical consultants see J. L. Heiberg, *Geschichte der Mathematik und Naturwissenschaften im Altertum*, pp. 47-49. B. L. van der Waerden, *Science Awakening*, 1st ed., pp. 277-78, refers to the collections of simple "Heronian" formulas as "cookbooks" and suggests that their disappearance is not to be mourned. See also Sir Thomas Heath, *A History of Greek Mathematics*, II, 307-8.

their classical origins[2] to their survivals in the late Middle Ages—is to appreciate the true character of Roman science and, at the same time, to discern one of the main causes of the low level of science in the Latin West before the Greco-Arabic revival in the twelfth century. Once the body of Latin scientific writing had been digested and codified in the age of Julius Caesar and Augustus, it did not change appreciably. The fatal error made by the Romans was in epitomizing Greek popular science, culling definitions and classifications while omitting proofs and analysis, and failing to comprehend the need for logical order and systematic development in a scientific discipline. The Carolingian commentators of Martianus Capella[3] would have been better able to interpret the science of Varro (116-27 B.C.) than Martianus had been to interpret the science of his Greek predecessors.

Because the texts of Latin science are more abundant from the early Middle Ages than from the classical period, greater attention must be paid than heretofore to the medieval texts in order to reconstruct Roman science. The quadrivium books of Martianus Capella's *De nuptiis Philologiae et Mercurii* are about as representative an example of Roman and medieval Latin science as is to be found in the extant literature. The disappearance of Varro's *Disciplinarum Libri IX* in the early Middle Ages leaves gaps in our knowledge of the Roman mathematical disciplines. Varro's work appears to have been the fountainhead of much of the subsequent Latin scientific literature. The gleanings from it in Aulus Gellius, Censorinus, Augustine, Cassiodorus, and

[2] A lucid account of the Greek quadrivium sciences is to be found in François Lasserre, *The Birth of Mathematics in the Age of Plato*. Lasserre points out that the content of the quadrivium was formulated by mathematicians who were contemporaries of Plato.

[3] John Scot Eriugena, Remigius of Auxerre, and Martin of Laon, all of the ninth century. The commentaries of the first two (see Bibliog.) have been edited by Cora E. Lutz in 1939 and 1962-1965, respectively. The commentary of Martin of Laon is being edited in part and, it is hoped, ultimately in full, by Jean Préaux, who made the important discovery that glosses contained in several ninth-century manuscripts of Martianus and formerly attributed to Dunchad were actually composed by Martin. Miss Lutz published the so-called Dunchad glosses, for Books II, IV, and V, in Dunchad *Glossae in Martianum*. Préaux states his reasons for attributing the glosses to Martin of Laon in "Le Commentaire de Martin de Laon sur l'œuvre de Martianus Capella," *Latomus*, XII (1953), 437-59.

Isidore indicate that it was a more refined and sophisticated version of Greek popular writings on the mathematical disciplines than the quadrivium part of Martianus' book, which was inspired by it. But, in the absence of Varro's work, Martianus' fifth-century version of the Latin quadrivium can give us a fair idea of the science of the Roman schools of the late Republic and early Empire, the science that had been available to Cicero, Vergil, and Ovid, and the science that was to prevail for more than a millennium in the Latin West, reaching its culmination among the twelfth-century Scholastics of Chartres.[4] Servius in the fourth century was able to comment intelligently and familiarly on the science of Vergil,[5] as was Remigius of Auxerre in the ninth century on the science of Martianus, because the content of their science remained static. Both commentators were schoolmasters and grammarians, rather than scientists, and they reflected the character of Latin science from its beginning—a subject for the school textbooks of grammarians and rhetors, and for polished gentlemen who wanted a smattering of it for a better grasp of literature and philosophy.

Martianus himself was such a gentleman, living in an age when the victory of Christianity over paganism was not yet complete.[6] Longstanding rivalries between Christians and pagans, and the more recent successes of Christianity, had intensified the desire of pagans to undertake, as a social responsibility, the preservation of classical culture. Cicero's ideal of the statesman, depicted, at the close of his *Republic*,[7]

[4] The profound influence that Martianus had upon the scholars at Chartres is seen in the recent study of Édouard Jeauneau ("Note sur l'École de Chartres," *SM*, ser. 3, V [1964], 821-65, esp. 842-43. See also Raymond Klibansky, "The School of Chartres," in *Twelfth Century Europe and the Foundations of Modern Society*, ed. by Marshall Clagett, Gaines Post, and Robert Reynolds (Madison, Wis., 1961), pp. 3-14.

[5] Attention should be drawn to the painstaking scholarship of Henry Nettleship in elucidating the character of the late Latin compiler literature in his *Lectures and Essays on Subjects Connected with Latin Literature and Scholarship*, and particularly to his essay on "Thilo's Servius," pp. 322-40.

[6] Important dates in the struggle are 324 (the position of the Christian Church was strengthened when Constantine I became sole emperor), 361-363 (the Emperor Julian tried to restore pagan religions), 384 (Symmachus unsuccessfully attempted to restore the Altar of Victory to Senate House), and 392 (Theodosius forbade pagan worship).

[7] In the episode known as *The Dream of Scipio*.

as a man equally devoted to philosophy and politics, finally found fulfillment in the littérateurs of the waning Empire. Many of the most eminent men of pagan letters of that time were also eminent politicians. A contemporary picture of such a company of savants and politicans–not all of them actually contemporaneous–is given by Macrobius in the opening book of his *Saturnalia*.[8]

Christians, too, had a pagan schooling and appreciated the importance of the secular disciplines to a zealous practitioner who would glorify God through reading, writing, and teaching or preaching. There is a sizable list of authors, either known or presumed to be Christians, who wrote upon subjects of pagan learning without making reference to Christianity.[9] A generation or two ago such an omission used to be offered as an argument that the author was not a Christian or at least not a Christian at the time of writing; but there is now a growing awareness that a writer on pagan subjects in the late Empire was almost exclusively a compiler of transmitted materials, who should not be expected to make reference to conditions of his own time.

The two Church Fathers who were most influential in cultivating an interest in pagan learning among Christians were Augustine and

[8] Macrobius himself admits at the outset (*Saturnalia* 1. 1. 5) that not all the guests at the Saturnalia banquet actually belong to the same age. A translation of Macrobius' *Saturnalia* by Percival V. Davies has recently appeared in the "Records of Civilization" (Columbia University Press, 1969). On the passionate determination of the pagan aristocracy to preserve classical literature and learning see Samuel Dill, *Roman Society in the Last Century of the Western Empire*, pp. 28-33, 154-58, and his summary statement (p. 391): "History shows few [other] examples of an aristocracy more devoted to letters than to war or sport or politics." See also T. R. Glover, *Life and Letters in the Fourth Century*, pp. 148-53.

[9] See Macrobius *Commentary on the Dream of Scipio*, tr. W. H. Stahl, pp. 7-9. M. Schanz and C. Hosius, *Geschichte der römischen Literatur* (hereafter cited as Schanz), Vol. IV, pt. 2, p. 316, n. 5, lists some writers, including Martianus, who make no mention of Christianity. Maïeul Cappuyns ("Capella [Martianus]," in *DHGE*, Vol. XI, col. 838) considers the possibility that Martianus was a Christian, but assumes at the same time that Macrobius and Boethius were Christians. H. Parker's suggestion ("The Seven Liberal Arts," *English Historical Review*, V [1890], 450) that Martianus' words *Syri cuiusdam dogmate* (142) may be a reference to Christianity is too precarious to be taken seriously. R. Turcan, "Martianus Capella et Jamblique,' *Revue des études latines*, XXXVI (1958), 239, takes the Syrian mentioned to be Iamblichus.

Cassiodorus. Augustine tells us[10] that he felt a desire to attain to spiritual truths through secular knowledge and that, while awaiting baptism at Milan, he occupied himself with the composition of manuals on the seven disciplines. The project was not carried very far, but the partial remains of his writings on the disciplines had a significant influence upon later Christian authors.[11] Cassiodorus did complete a book on the secular disciplines (*Institutiones II*) to complement his book on spiritual learning (*Institutiones I*), and, significantly enough, it was Book II that had the wider circulation in manuscripts and exercised the greater influence upon intellectual developments in the West. The importance of Cassiodorus in keeping alive an interest in classical learning and in the ultimate preservation of the manuscript texts of classical Latin authors can hardly be overemphasized.[12]

We may deplore the state of classical learning–science and philosophy in particular– during the early Middle Ages, as judged by the level of the compilations of the pagan savants and Church Fathers of the fourth, fifth, and sixth centuries. But it is important to remember that these men in their zeal preserved at least the scraps of classical learning and inspired scribes in the centuries following to transmit copies of their compilations and of the texts of selected classical Latin authors to scholars of the later Middle Ages. The opinion of F. W.

[10] *Retractiones* I. 6.

[11] He completed a book on grammar (*De grammatica*) and a large portion of a work on music (*De musica*). He tells us that his notes on the other five disciplines were lost. He was planning to deviate from Varro's canon by substituting philosophy, a subject acceptable to theologians, for astronomy, suspect because of its associations with astrology, and by omitting Varro's disciplines of medicine and architecture. On the extant remains of Augustine's manuals see Marrou, *Saint Augustin*, pp. 570-79. On Augustine's secular background see David Knowles, *The Evolution of Medieval Thought*, chap. 3.

[12] It was Cassiodorus who convinced the clergy that the pagan classics had a rightful place in Christian education. M. R. James, in *CMH*, III, 486, calls Cassiodorus "the greatest individual contributor to the preservation of learning in the West" and says that but for him it is quite possible that no Latin classic except the works of Vergil would have come down to us in complete form. A recent study of the seven liberal arts, by Friedmar Kühnert, *Allgemeinbildung und Fachbildung in der Antike*, stresses the importance of Martianus, Cassiodorus, and Isidore. Kühnert cites the references to the liberal arts in ancient authors and provides a comprehensive bibliography of modern studies.

Hall, an expert on the transmission of classical texts in later ages, is worth quoting here: "What has preserved the Latin classics for us is not the Roman libraries, but the efforts of pagan nobles of the Theodosian epoch—the 'anti-Christian Fronde,' as they have been called. These men kept alive the ancient learning long enough to breed up men of the type of Cassiodorus in the place of early fanatics.'[13] Without the texts of these pagan nobles intellectual contacts with the classical world would have been broken and the Dark Ages would have been much darker. Such specimens of classical learning as did survive instilled in scholars like Gerbert and Adelard a craving for better examples of classical science and philosophy—the original Greek works that were circulating in the Arabic world and that the Romans a thousand years earlier had neglected to transmit to the Western world. The Greco-Arabic revival in the twelfth century saw the translation into Latin of major scientific works of Hippocrates, Euclid, Aristotle, Archimedes, Apollonius of Perga, Ptolemy, Galen, and others; and that revival was followed by the Renaissance. There would have been no classical Latin revival in the Renaissance, and almost none of our classical Latin authors would have survived if the cultivated pagans and Christians of late antiquity had not been devoted to letters as well as to politics and religion.[14]

[13] *Classical Review*, XXXVI (1922), 32.

[14] The author of the article "Scholasticism" in the *Encyclopaedia Britannica*, 11th ed., enumerates the following works as the sum total of philosophical writings available to schoolmen of the early Middle Ages: Boethius' logical treatises and commentaries and his Latin versions of the *De interpretatione* and *Categoriae* of Aristotle and of Porphyry's *Isagoge*, Calcidius' translation of and commentary on Plato's *Timaeus*, Apuleius' *De dogmate Platonis*, Macrobius' commentary on Cicero's *Somnium Scipionis*, Augustine's writings, Martianus Capella's *De nuptiis Philologiae et Mercurii*, Cassiodorus' *Institutiones*, and the *Etymologiae* of Isidore of Seville.

The Author

VERY LITTLE is known for certain about Martianus Capella's life. Nearly all of our information comes from his book, and the interpretations given to his autobiographical statements are for the most part insecure, because of the ambiguity and obscurity of his style and the corrupt condition of the manuscripts. Some inferences, more or less bold, may be drawn from the body of his work. That he did not belong to the elite we may surmise from his debased style and from his lament about being impoverished and settling in old age in a neighborhood of slothful oxherds.[1] The common assumption that he was a high-ranking politician rests upon a bit of tenuous evidence—a single line in the badly corrupted text of the autobiographical poem that concludes his work, a line that is open to quite different interpretations, as we shall see.

Martianus' range of learning did not include a serious interest in the philosophers, for whom he expressed mild contempt.[2] Among the more than one hundred guests at the heavenly wedding of Mercury and Philology, most of the lesser deities or personifications, we meet some of the most illustrious of Greek philosophers—Plato, Aristotle, Thales, Pythagoras, Heraclitus, Democritus, Epicurus, Zeno, Arcesilaus, Carneades, and Chrysippus. But to Martianus these were mere names, introduced to give his book an air of authority.[3] When the philosophers are cited or called upon to demonstrate their wisdom, the matter under discussion is not some metaphysical doctrine but merely

[1] 999 (Dick 534. 14-15). (The citations for Martianus refer first to the section numbers indicated in the margins of modern editions and then to the page and lines in the Dick edition.)

[2] He calls them "starveling and unkempt": 578 (Dick 288. 6-10); Remigius *ad loc.* (ed. Lutz, II, 129) identifies the philosophers described by Martianus as Sophists, Stoics, and Cynics. Elsewhere Martianus says that they are "abstruse and ostentatious": 812 (Dick 429. 19-22) and Remigius *ad loc.* (ed. Lutz, II, 249-50).

[3] He also includes among the guests some mythical authorities on learned matters: Linus, Musaeus, Orpheus, and Amphion.

an elementary principle in one of the trivium or quadrivium disciplines.[4] Eminent professional mathematicians and astronomers are also included among the wedding guests—Euclid, Archimedes, Eratosthenes, Hipparchus, and Ptolemy—but again these figures serve as decorations.

That Martianus occasionally introduces Neoplatonic terminology and seems to be expressing Neoplatonist and Neopythagorean doctrines[5] must not be taken to indicate that he was a follower of the Neoplatonic school of philosophy. Neoplatonism was the only pagan philosophy to flourish in the last century of the Western Empire, and its adherents took a leading part in the bitter conflict with Christianity.[6] The remnants of secular philosophy and scientific learning that survived were largely in the Platonic tradition, stemming ultimately from Plato's *Timaeus*.[7] From the time of its composition until the late Middle Ages, that book inspired generations of commentators on, and popularizers of, works on theoretical cosmography and arithmology, and it is not to be expected that a Latin compiler of traditional and conflated doctrines on the quadrivium would wholly avoid the use of Neoplatonic vocabulary.[8]

[4] Aristotle, Carneades, and Chrysippus on dialectic; Pythagoras on arithmetic and astronomy; and Plato on astronomy.

[5] 92 (Dick 39. 15); 126 (Dick 56. 18); 185 (Dick 73. 10-16); 203 (Dick 77. 3-4); 567 (Dick 285. 11); 922 (Dick 490. 11; 14-19).

[6] See Dill, pp. 100-6; Glover, pp. 63-65, 184-93, 211-13; W. H. V. Reade, "Philosophy in the Middle Ages," in *CMH*, V, 781.

[7] Paul Shorey wrote (*Platonism Ancient and Modern* [Berkeley, 1938], p. 105): "The shortest cut to the study of philosophy of the early Middle Ages is to commit the *Timaeus* to memory. Otherwise you can never be sure that any sentence that strikes your attention is not a latent quotation from the *Timaeus*, or a development of one of its suggestions." As this statement reflects, Professor Shorey was a staunch supporter of Plato; but he was also a careful and respected scholar, and his view has much to commend it. Reade, pp. 789-90, also stresses the key position of the *Timaeus*. For instances of *Timaeus* influence in the traditions of ancient and medieval science see Stahl, *Roman Science*, pp. 22, 47-48, 55-56, 68, 121, 128-29, 142.

[8] Turcan, pp. 235-54, sees in the precise terminology used by Martianus in describing the hierarchy of deities and the planetary demarcations in Philology's ascent in the latter part of Book II definite traces of borrowing from the Neoplatonist Iamblichus. It was the considered opinion of Pierre Courcelle, (*Les Lettres grecques en occident de Macrobe à Cassiodore*, p. 203, n. 8), however, that

Differences of opinion have arisen about most of the details and circumstances of Martianus' life.[9] He referred to his book as *senilem fabulam*, a tale composed in his declining years,[10] and addressed it to his son Martianus.[11] The author's full name is given in the subscriptions of various manuscripts as Martianus Min(n)e(i)us Felix Capella.[12] He refers to himself as Felix (576) and Felix Capella (806, 999). He is called Felix Capella by Fulgentius[13] and Cassiodorus.[14] Gregory of Tours, in a fine appreciation, calls him "our Martianus."[15] He is called Martianus by the ninth-century commentators of his book, John Scot Eriugena and Remigius of Auxerre,[16] and at the present time, if a

---

Iamblichus was not read by the Latin writers of the West, and Courcelle's opinion was based on more than a single statement of a Carolingian monk, as Turcan (p. 236) supposes. See Courcelle, p. 394.

[9] For other discussions of the meager and often ambiguous information provided by Martianus and the manuscripts, see Claudio Leonardi, "I codici di Marziano Capella," *Aevum*, XXXIII (1959), 443-44; Cappuyns, "Capella," cols. 836-38, 842-43; Paul Wessner, "Martianus Capella," in Pauly-Wissowa, Vol. XIV (1930), cols. 2003-4; Dick, pp. xxv-xxvi; Schanz, Vol. IV, pt. 2, pp. 168-68; W. S. Teuffel and L. Schwabe, *History of Roman Literature* (hereafter cited as Teuffel), II, 446-47; *Martianus Capella*, ed. F. Eyssenhardt, pp. iii-ix.

[10] 997 (Dick 533. 11). At sections 2 (Dick 4. 6-8) and 1000 (Dick 535. 4-5) he refers to his composition as the "silly trifles" of an "old man" and begs his son's indulgence. Some scholars (including Lewis and Short, *A Latin Dictionary: s.v.* "decurio") have taken Martianus' expression *incrementisque lustralibus decuriatum* (2; Dick 4. 7-8) to mean that he was in his early fifties (a man of ten lustra). Such an interpretation would not agree with his references to his senility in the other passages cited, however; and elsewhere (728) Martianus uses *decuriatus* to mean "a decad of numbers." Remigius (Lutz, I, 70) and John Scot Eriugena (*Annotationes in Marcianum*, ed. Lutz, p. 5) interpret the words as referring to his advanced age and legal distinctions. R. E. Latham, ed., *Revised Medieval Latin Word-List*, cites the use of *decuriare* in the fifteenth century with the meaning "to put out of court." If Martianus spent much of his time in legal practice, as appears to have been the case, a likely translation of *decuriatus* would be "finished with the courts."

[11] 2 (Dick 4. 8); 997 (Dick 533. 11).

[12] Leonardi, "I codici" (1959), p. 443.

[13] *Expositio sermonum antiquorum* 45 (ed. R. Helm, p. 123).

[14] *Institutiones* 2. 2. 17; 2. 3. 20.

[15] *Historia Francorum* 10. 31 (ed. Arndt, p. 449; Migne, *PL*, LXXI, 572).

[16] Remigius calls him Martianus Capella in the title of his commentary, Martianus in the frequent references to him in the text.

single name is used, Martianus seems to be preferred to Capella.

Martianus may have been born at Carthage; that he grew up and lived most of his life there is attested by the words *Afer Carthaginiensis* [African of Carthage] added to his name in the manuscripts and by his reference to himself as a "fosterling of the prosperous city of Dido."[17] Two mistaken notions about his locale have persisted to the present: that he was born at Madaura, about 150 miles southwest of Carthage, an assumption for which there is no evidence;[18] and that he spent his old age at Rome, an equally unfounded assumption.[19]

The two matters about which we would most like to feel assured are the very ones in greatest dispute—the date of composition of Martianus' book and the details of his professional career. Recent scholars generally have accepted a date between Alaric's sack of Rome in 410 and the Vandal Gaiseric's crossing to North African shores in 429 or his occupation of Carthage in 439.[20] The two most recent editors of Martianus' work inclined to an earlier date. Eyssenhardt, in a doctoral dissertation published in 1861,[21] places Martianus between 284 and 330,

[17] 999 (Dick 534. 13): *beata alumnum urbs Elissae quem vidit.*

[18] Cappuyns, "Capella," col. 837, traces the notion to a statement in the Grotius edition (1599) which was later given currency by Fabricius in his *Bibliotheca Latina* (1697). The most recent scholar to repeat the error is F. J. E. Raby, *A History of Secular Latin Poetry in the Middle Ages*, I, 100.

[19] In the closing autobiographical poem Martianus speaks of himself "as if just coming from the Court of Mars" (*utque e Martis curia*); but he is here borrowing a phrase, as is his wont, from a classical Latin poet (Juvenal *Satires* 9. 101: *ut curia Martis Athenis*). Juvenal is referring not to the Roman Curia but to the Areopagus Court at Athens. Remigius *ad loc.* (ed. Lutz, II, 368) assumes that Martianus is referring to the Areopagus. Later in the poem Martianus speaks of his abode as *iugariorum murcidam viciniam*, literally, "the slothful neighborhood of oxherds." Some have taken this to refer to the *Vicus Iugarius*, a street in Rome, though the manuscripts all read *viciniam* or *vicinam*. On the *Vicus Iugarius* see S. B. Platner and T. Ashby, *A Topographical Dictionary of Ancient Rome* (Oxford, 1929), p. 574.

[20] Leonardi, "I codici" (1959), p. 443; James Willis, *Martianus Capella and His Early Commentators*, pp. 6-8; Cappuyns, "Capella," col. 838; Courcelle, *Lettres grecques*, p. 198; E. S. Duckett, *Latin Writers of the Fifth Century*, p. 224; Schanz, Vol. IV, pt. 2, p. 169; J. E. Sandys, *A History of Classical Scholarship* (Cambridge, 1906), I, 241-42; Paul Monceaux, *Les Africains: Étude sur la littérature latine d'Afrique*, p. 445; Teuffel, II, 447.

[21] *Commentationis criticae de Marciano Capella particula*, pp. 14-15.

setting the *terminus ante quem* at 330 because Martianus uses the name Byzantium for the city which Constantine made the capital of his Eastern Empire and renamed Constantinople in 330. Eyssenhardt was unaware that it was standard practice for a compiler to copy information from his predecessors even when he knew that their information was outdated.[22] Martianus copied his statement about Byzantium almost verbatim from Solinus, who in turn had paraphrased the statement he found in Pliny.[23] In his edition of Martianus, published five years later, Eyssenhardt became more cautious about Martianus' dates, placing firm reliance only upon a *terminus ante quem* of 439 and wondering how Martianus could have been so stupid as not to change the name of Byzantium if he had written after 330.[24] Dick, the most recent editor, accepted Eyssenhardt's earlier termini of 284 and 330[25] and supported Eyssenhardt's contentions by pointing to Martianus' care and skill in handling many meters, as noted by Stange,[26] and supposing that such metrical skills would not be found in a writer of the fifth century. But, as Willis observes, metrical skills vary greatly among poets, and there is no reason to deny such skills to a writer familiar as Martianus was with the classical Latin models simply because his contemporaries were not as skillful as he.[27] Paul Wessner, author of the "Martianus" article in the Pauly-Wissowa *Real-Encyclopädie der classischen Altertumswissenschaft* (1930),[28] regarded the

[22] Pliny, for example, in the table of contents listed at the opening of his *Natural History*, acknowledges that he has included in Book VI ninety-five extinct towns and peoples among his geographical names.

[23] Martianus 657 (Dick 325. 12-13): *promuntorium Ceras Chryseon Byzantio oppido celebratum.* Solinus 10. 17: *promuntorium Ceras Chryseon Byzantio oppido nobile.* Pliny 4. 46: *promunturium Chryseon Ceras in quo oppidum Byzantium.* It is noteworthy that Theodor Mommsen, in his masterly edition of Solinus' *Collectanea rerum memorabilium*, p. vi, inclined to date Solinus before 330 because of this very passage, although he admitted that a statement repeated by a compiler does not have a decisive bearing. On the confusions and errors resulting from Martianus' careless habits of borrowing and excerpting from his predecessors see Stahl, *Roman Sciences*, pp. 176-77, 279-80.

[24] *Martianus*, ed. Eyssenhardt, pp. vii-ix.

[25] Dick, p. xxv.

[26] F. O. Stange, *De re metrica Martiani Capellae*, p. 38.

[27] Willis, p. 9.

[28] Vol. XIV, col. 2004.

arguments of Eyssenhardt and Dick as cogent and added a flimsy argument of his own, drawn from a casual remark of Ammianus Marcellinus (*c.* 385).[29] Marcellinus mentions that some of his contemporaries read only Juvenal and Marius Maximus; Martianus is found quoting Juvenal five times in his book and therefore, according to Wessner, belongs to the time of Ammianus! Such is the case for assigning Martianus to the fourth or third century.

The prevailing view, that Martianus wrote his book after Alaric's sack of Rome in 410, is based upon this passage in Book VI: *Ostia Tiberina dehicque ipsa caput gentium Roma*, armis, viris, sacrisque, quamdiu viguit, caeliferis laudibus conferenda (37; Dick 311. 11). As it is usually interpreted,[30] this passage indicates that Rome was at that time no longer in her prime. Here too Martianus is drawing upon Pliny: *Tiberina ostia et Roma terrarum caput* (*Natural History* 3. 38); but the unitalicized words in Martianus' statement are his own and reflect his own feelings about the grandeur of the Rome that was. Taking *quamdiu viguit* to mean "all the time she has been in her prime," as Dick and others who assign instead Martianus to a period before the fall propose–instead of the usually accepted meaning, "as long as she was in her prime"– is a forced interpretation.[31] On the other hand, placing all the weight of a *terminus post quem* on this expression of a florid writer is admittedly not secure, since the phrase does not necessarily imply that Rome was vigorous right up to the fall.

The argument that the book was writen before 429 or, at the latest, 439 is based upon this phrase of Martianus: *Carthago inclita pridem armis, nunc felicitate reverenda* [Carthage, once famed for her military prowess, now awesome in her prosperity (669; Dick 333. 5-6)] and on the probable mention of the high office of proconsul in a corrupt line in the poem that closes his work: *proconsulari vero dantem*

[29] *Res gestae* 28. 4. 14.

[30] "Rome herself, as long as she was in her prime, a city to be extolled to the skies for her military prowess, men, and religious worship. . . ."

[31] Dick, p. xxv, explains *quamdiu viguit* as meaning *ab originibus usque ad nostra tempora*, i.e., *omni tempore*. Courcelle, *Lettres grecques*, p. 198, does not accept this interpretation; nor does Willis, p. 8, who states that Dick's translation of *quamdiu* viguit is "very dubious Latinity" and that if Rome was still the capital in Martianus' day, there would have been no reason to add his own qualifying statement to the one he found in Pliny.

culmini (999; Dick 534. 10). Only one scholar, to my knowledge, does not interpret *proconsulari culmini* as referring to the office of proconsul.[32] Recent scholars, assuming that the office was abolished with Gaiseric's capture of the city in 439, agree that Martianus was writing before that date.[33] (This assumption is altogether likely, though not absolutely certain.[34]) Moreover, Martianus' descriptive remark about happy conditions in Carthage is a personal one, not drawn from his predecessors Pliny and Solinus, and could scarcely have applied to Carthage after the Vandal occupation.

The evidence from Martianus' own statements, then, points to a date of composition between 410 and 439, the terminus at 439 being more reliable than that at 410. To assign the work to a date after 439 calls for extravagant assumptions—either that the words *proconsulari culmini* are a topographical reference or that the title and office of proconsul continued in use during Gaiseric's reign at Carthage; and that Martianus could be describing conditions at Carthage under Gaiseric as prosperous and happy.[35]

Little reliance for dating can be placed on Martianus' style. Stylistic arguments for an earlier date, on the grounds that Martianus' metrical skills are found only in poets of an earlier period, can easily be offset by considerations of his prose style, a bizarre one quite unlike the styles of other Latin writers of his age and appearing to belong to a

[32] Parker, p. 442, assumes that the phrase is a topographical reference to a "proconsular ridge," in the city of Carthage. The difficulties of interpreting this line in its context as referring to Martianus as proconsul will be discussed below.

[33] A late date for the composition, around 470, was widely accepted by earlier scholars, beginning with G. J. Vossius in 1627. *Martianus Capella*, ed. F. Eyssenhardt, pp. iv-v, briefly discusses the opinions of some of these scholars.

[34] An expert on Roman provincial administration and law recently expressed to me the opinion that Gaiseric could have kept the title and office in use. Ludwig Schmidt (*CMH*, I, 307), however, points out that there is no trace of reckoning according to consular years or indictions at Carthage after 439, as there is, e.g., from the Burgundian Kingdom.

[35] There are a few Vandal coins bearing the inscription FELIX CARTHAGO. See Warwick Wroth, *Catalogue of the Coins of the Vandals, Ostrogoths and Lombards... in the British Museum* (London, 1911), p. 13, nos. 3-5 (dated 523-530); nos. 1-2, probably of Vandalic origin, were struck by Hilderic during the reign of Justin I (518-527). On contemporary accounts of the sack of Carthage see Courcelle, *Histoire littéraire des grandes invasions germaniques*, 3d ed., pp. 129-39.

later, rather than an earlier, period. The next *terminus ante quem* after 439 is found in a mention of Martianus in Fulgentius, who probably flourished at the close of the fifth century or the beginning of the sixth.

Martianus' career and occupation are matters of open speculation, ranging from Parker's opinion that he was an impecunious and self-taught peasant[36] to the rather bold supposition that he attained to the proconsulship.[37] The latter conjecture would place Martianus in the select company of pagan savant-politicians who felt a deep obligation to preserve secular learning in their writings. This is the most precarious of the conjectures about Martianus' career, all of which are drawn from a few lines in his closing poem—lines so corrupt that they are probably beyond hope of repair[38]—and from statements made in the setting portion of his work. Martianus several times expresses concern about tiring his readers with the subject matter of his seven arts and, to relieve the tedium, he generally ends each discourse with a flighty poem. The final, autobiographical poem, serving as a sort of peroration to the work and containing the musings of an author who ridiculed and belittled himself several times earlier, is shot through with flowery, figurative, and ambiguous diction. It is not surprising that the text of this poem is one of the most corrupt passages in the entire work. The lines in question, according to the reading adopted by Dick (534. 10-12), are

[36] Parker, pp. 442-44, reaches this conclusion from Martianus' poor style and his remark, at the close of his book, about living in poverty among ploughmen.

[37] Among those who have thought that Martianus was referring to himself as proconsul are Leonardi, "Intorno al 'Liber de numeris' di Isidoro di Siviglia," *BISIAM*, LXVIII (1956), 215: "probabilmente"; Leonardi, "I codici" (1959), p. 443: "forse . . . ebbe il proconsolato"; and Cappuyns, "Capella," cols. 836, 838. Dick seems to me to be equivocating in the reading he adopts. Perhaps the most weighty authority adopting this view is the ninth-century commentator Remigius, who was using a manuscript of Martianus older than any we possess. See n. 39.

[38] Jean Préaux informs me in a letter that he is preparing a new text of this poem, based upon a collation of the best manuscripts, and that he does not interpret the text to mean that Martianus held a proconsulship. Other noteworthy recent attempts to emend and interpret the text of the poem are those of Cappuyns, "Capella," col. 836; Dick, pp. 534-35; C. Morelli, "Quaestiones in Martianum Capellam," *SIFC*, XVII (1909), 247; Monceaux, pp. 445-47; and Parker, pp. 440-44.

† proconsulari vero dantem culmini
ipsoque dudum bombinat ore flosculo
† decerptum falce iam canescenti rota.

If the readings of the best manuscripts are retained, the syntax is faulty. Many conjectural readings have been offered to provide a syntactical and plausible interpretation. The readings *vero dantem* (10) and *bombinatorem* (11), found in several manuscripts, including the one used by Remigius in the ninth century,[39] imply that Martianus is referring to himself as a bumblebee or hummingbird cut off from his flower—an apt figure for himself—and that he is giving himself to the office of proconsul. Cappuyns adopts these readings and infers that Martianus held the proconsulship. But Kopp prefers the reading *perorantem* for *vero dantem*, and Eyssenhardt and Dick, while retaining the participle *dantem*, give it no object. Kopp's emendation of *perorantem*, and Sundermeyer's of *verba dantem*, in place of *vero dantem*, imply that Martianus was pleading cases before the proconsul of Africa.

The last implication is perhaps a safer one and is the one accepted by Willis, Monceaux, and Teuffel.[40] Curtius, Raby, Morelli, Beazley, and Kopp infer that he was a lawyer or advocate,[41] but it would be impossible to determine from Martianus' statements whether he was a lawyer, advocate, pleader, or merely a practitioner of the rhetorician's art.[42] He could have been pleading cases before a proconsul in any of these roles.[43] The technicalities of legal procedure and the

[39] Remigius (ed. Lutz, II, 369) explains *dantem* as *scribentem suas fabulas* and *bombinatorem* as referring to Martianus as *pompose loquentem*. Remigius assumes that Martianus held the proconsulship at that time: *Significat enim tunc illum proconsulem Carthaginis fuisse quando hunc librum scripsit*. Regarding a reading for *bombinator* the *Thesaurus Linguae Latinae* despairs: *omnia incerta*.

[40] Willis, p. 10; Monceaux, p. 445; Teuffel, II, 447.

[41] E. R. Curtius, *European Literature and the Latin Middle Ages*, p. 75; Raby, *Secular Latin Poetry*, I, 101; Morelli, p. 250; C. R. Beazley, *The Dawn of Modern Geography*, I, 341; Martianus *De nuptiis Philologiae et Mercurii et de septem artibus liberalibus libri novem*, ed. U. F. Kopp, pp. 488, 771.

[42] G. W. Clarke addresses himself to this very problem in the case of St. Cyprian: ("The Secular Profession of St. Cyprian of Carthage," *Latomus*, XXIV [1965], 633-38).

techniques of arguing cases get his first and major attention in his book on rhetoric.[44] His undertaking the laborious task of compiling a systematic handbook of the seven liberal arts suggests that at some time he was a teacher of rhetoric. Some scholars have also referred to him as a grammarian.

The inference that he practiced law has been drawn from two autobiographical passages, both written in flowery and ambiguous language. Referring to himself by name in the closing poem, Martianus goes on to say: *indocta rabidum quem videre saecula iurgis caninos blateratus pendere* (999; Dick 534. 8-9). *Iurgis* is generally taken to mean "legal disputes"[45] by those who assume that Martianus was involved in legal cases. The meaning of the passage is far from certain, however. The manuscript reading of the text used by Remigius in the ninth century was *lurgis*, which he glossed as *improbis, gulosis, epulonibus, hiatibus*; the text of John Scot Eriugena had the same reading and he glossed the word as *gluttosis*.[46] Remigius took *pendere* to mean *impendere* or *reddere*, as if to say that Martianus was paying back the barkings of dogs to gluttonous men.[47]

In the second passage Satire is upbraiding Martianus because his strenuous exertions and devotions to the wrangling and pettifoggery of the courts[48] have tied him down and blunted his edge for a better

[43] My informant on Roman law in the provinces says that there is no evidence that cases were pleaded in the late Empire in the same manner as before A.D. 250, except in a few capitals like Byzantium and Alexandria. Anyone could inform himself by reading the legal tracts circulating at the time and could undertake to advise the magistrate who was charged with making a decision in civil cases. For a bibliography on legal practice in the Roman Empire see N. Lewis and M. Reinhold, *Roman Civilization, Sourcebook II: The Empire* (New York, 1966), pp. 633-34.

[44] Book V: 441, 443-75, 498-503, 553-65.

[45] A possible translation would then be: "... whom ignorant generations have seen ranting away and weighing dogs' barkings in legal cases."

[46] *Annotationes*, ed. Lutz, p. 220.

[47] Parker, p. 442, unaware of the readings and glosses of Remigius and John Scot translated the passage as "who has, under the eyes of an ignorant generation, paid back in anger with abuse the yelping of dogs."

[48] This is the usual interpretation of the phrase, as referring to the pleading of cases. See Martianus *De nuptiis*, ed. Kopp, p. 488; *Martianus*, ed. Eyssenhardt, p. v; Wessner, col. 2004.

occupation, namely, the compilation of a learned book: *desudatio curaque districtior tibi forensis rabulationis partibus illigata aciem industriae melioris obtudit* (577; Dick 287. 19-21). But the words *forensis rabulationis* are ambiguous; they can mean either "wrangling in the market place" or "forensic pettifogging."[49] Parker,[50] who thinks Martianus was a farmer, interprets the expression as "disputes in the market." John Scot[51] and Remigius[52] both understood the term as "rhetorical disputes."

No one has yet drawn attention to Martianus' occasional use of technical and quasi-legal vocabulary—for example, in the discussion of the dowry for the wedding. The seven ladies who present their disciplines before the wedding ceremony are called *feminae dotales*.[53] The bride's mother requests[54] that at the time the dowry is formally presented, there be a reading of the *Lex Pappia Poppaea*, a law dealing with marriage contracts, which Augustus promulgated in A.D. 9. Near the close of *The Marriage*, after six of the seven bridesmaids have presented their lengthy discourses and the hour is getting late, it is pointed out that seven other maidens, representing the seven prophetic arts, are standing by and might be added to the dowry.[55] Apollo makes a strong case for hearing them, but Jupiter is eager to bring the wedding ceremony to an end. Luna suggests an adjournment to provide the guests with a respite and a later opportunity to hear the remaining learned maidens. A technical question is then raised in which legal

[49] Lewis and Short's *Latin Dictionary* applies double meanings to each word, and to *forensis* in a passage in which Quintilian (*Institutio oratoria* 5. 10. 27) is speaking of men in different walks of life, the dictionary assigns the additional meanings "public pleader, advocate." Firmicus Maternus' remarks in the proem to Book IV of his *Matheseos libri VIII*, about the rantings of the law courts (*caninae contentionis iurgiosa certamina*) and about his feelings of relief upon his retirement from the legal practice (*in otio constitutus et forensium certaminum depugnationibus liberatus*) in order to begin the composition of a book on astrology, are strikingly like the feelings and expressions of Martianus.

[50] P. 443.

[51] *Annotationes*, ed. Lutz, p. 135.

[52] Remigius, *ad loc.* (ed. Lutz, II, 128).

[53] "Ladies of the dowry." See 803 (Dick 422. 12; 423. 3); 807 (Dick 426. 6); 810 (Dick 428. 9).

[54] 217 (Dick 79. 9-14).

[55] 892; Dick 472. 7-10.

vocabulary (not italicized) is used:[56] *utrum repensatrix data diesque conferendae dotis prorogari iure publico posset inquiritur. quo dicto arcanus ille prisci iuris assertor magna nepotum obsecratione* consulitur respondit*que regulariter etiam matrimonio copulato* dotem dicere *feminam viro* nullis legibus prohiberi.[57]

[56] 898; Dick 475. 19–476. 5. A professor of Roman law informs me that the unitalicized words are legal vocabulary, but he is of the opinion that the passage could have been written by a layman.

[57] Literally translated: "whether a recompense could be offered and the day for conferring the dowry could be postponed, according to public law. At this suggestion that cryptic authority on ancient law [Saturn] was besought by the entreaties of his numerous descendants and he replied that, according to custom, once the couple had been joined in marriage there was no statutory objection to the wife's constituting the dowry for her husband."

The Work

ONE CANNOT READ Martianus' book cursorily—one must tackle it—and the reader is immediately at a loss to explain how a book so dull and difficult could have been one of the most popular books of Western Europe for nearly a thousand years. We moderns may be repelled by the style and content of the *De nuptiis*, but vernacular readers and medieval students seeking an introduction to the learned arts and finding in Martianus' work a fairly compact treatise dressed in fantasy and allegory were both charmed and edified by it. Martianus understood the tastes of his readers much better than the modern critics who have been puzzled by this apparent enigma. He was himself apologetic about introducing banter and cheap fiction into a serious discourse on learned subjects and he frequently represents himself as a silly old man. However, his self-chastisement was not so caustic as the remarks of modern critics.[1]

The *De nuptiis Philologiae et Mercurii*,[2] according to H. O. Taylor,

[1] C. S. Lewis, *The Allegory of Love*, p. 78, remarks: "For this universe, which has produced the bee-orchid and the giraffe, has produced nothing stranger than Martianus Capella." Dill, who read the literature of this period with great care and insight, aptly observes (p. 412): "It is difficult to conceive the state of culture where the mixture of dry traditional school learning and tasteless and extravagant mythological ornament, applied to the most incongruous material, with an absolutely bizarre effect, could have been applauded as a sweetener of the toils of learning." H. J. Rose, *A Handbook of Latin Literature* (1st ed., New York, 1936; repr., 1960), p. 458, is less tolerant: "It is the dullest and poorest stuff imaginable." Cora Lutz has collected other adverse comments in "Remigius' Ideas on the Origin of the Seven Liberal Arts," *Medievalia et humanistica*, X (1956), 32-33. For other unfavorable comments see Leonardi, "Nota introduttiva per un'indagine sulla fortuna di Marziano Capella nel medioevo," *BISIAM*, LVII (1955), 269, n. 1.

[2] The title is uncertain. Fulgentius, not very long after the work was written, referred to it as *Liber de nuptiis Mercurii e Philologiae*, but this title in the manuscripts refers to only the first two books, the setting of the heavenly marriage. The remaining books are titled according to the disciplines presented: *De arte grammatica, De arte dialectica*, etc. Cassiodorus (*Institutiones* 2. 2. 17) calls the

was "perhaps the most widely used schoolbook of the Middle Ages."[3] It would be hard to name a more popular textbook for Latin readers of later ages. It had to withstand keen competition from Boethius, Cassiodorus, and Isidore of Seville, but it had the salient advantage of offering a well-proportioned and comprehensive treatment of all the liberal arts in the compass of one comfortable-sized book. The *De nuptiis* was the foundation of the medieval trivium and quadrivium.[4] Since it recapitulated the fundamentals of the Roman academic curriculum and transmitted them to later generations of students, the book must be regarded as the key work in the history of education, rhetoric, and science[5] during this period. It therefore ill behooves critics to ridicule and deplore Martianus' book, the loss of which might have had a notably adverse effect upon intellectual life in Western Europe, such as it

---

book *De septem disciplinis* but admits (2. 3. 20) that he has not seen a copy of it. Martianus may have called the book *Disciplinae*. See Schanz, Vol. IV, pt. 2, p. 169; Wessner, cols. 2004-5; Cappuyns, "Capella," cols. 838-39. The title *Satyricon*, once widely used, is no longer current.

[3] *The Classical Heritage of the Middle Ages* (New York, 1901), p. 49. C. H. Haskins, *The Renaissance of the Twelfth Century*, p. 81, thinks it somewhat exaggerated to call Martianus "the most popular writer in the Middle Ages after the Bible and Vergil."

[4] P. O. Kristeller traces the development of the arts, and points to Martianus' significant role, in a delightful and impressively erudite essay, "The Modern System of the Arts," reprinted in his *Renaissance Thought II*, pp. 163-227. See esp. pp. 172-74. In a sequel to the studies in *Renaissance Thought*, entitled *Renaissance Philosophy and the Mediaeval Tradition* (p. 45), Professor Kristeller makes the interesting observation that only two groups of sciences, medicine and the mathematical disciplines, had a separate history from antiquity to the Renaissance that was "relatively, if not entirely, independent of philosophy." For a survey of the liberal arts in Western education see R. M. Martin, "Arts Libéraux (Sept)," in *DHGE*, Vol. IV (1930), cols. 827-43.

[5] A few prominent authorities on the history of medieval science who call attention to the importance of Martianus are P. Duhem, *Le système du monde*, II, 411; III, 62, 110; Charles Singer, *A Short History of Scientific Ideas to 1900* (Oxford, 1959), p. 138; A. C. Crombie, *Medieval and Early Modern Science* (New York, 1959), I, 14 (but note his classification of Martianus' book as a Greek commentary); *Ancient and Medieval Science*, ed. René Taton (New York, 1963), p. 367; C. H. Haskins, *Studies in the History of Mediaeval Science*, p. 89; Lynn Thorndike, *A History of Magic and Experimental Science*, I, 545; J. L. E. Dreyer, *A History of Astronomy from Thales to Kepler*, p. 207; J. K. Wright, *Geographical Lore of the Time of the Crusades*, pp. 9, 11.

was, and upon the chances of a revival of learning in the later Middle Ages. W. P. Ker, in a book that has become a classic on this subject, flatly states: "If Martianus Capella had been forgotten, with the school traditions exemplified in his book, there would have been no chance of a revival of learning"[6]—an arresting testimonial to Martianus' importance but at the same time a patent overstatement of the case.[7]

Another proof that Martianus correctly understood the tastes of his readers is seen in the influence exerted on genre-writing by the setting portions of his book. *The Marriage of Philology and Mercury* and Boethius' *Consolation of Philosophy* were the two main prototypes of the prose-verse, chantefable form of fiction throughout the Middle Ages.[8] In allegory Martianus shared the dominant position with Prudentius, Martianus having the greatest influence on secular, Prudentius on Christian, allegory. Moreover, the celestial marriage of Mercury and Philology inspired concepts and accounts of heavenly journeys in the later Middle Ages, up to and including Dante. Equally important was the influence that Martianus' allegorical figures had upon medieval art (see Appendix A, which traces the later fortunes of his seven learned bridesmaids in sculpture and painting).

The allegorical setting, occupying the first two books, was a delight to medieval readers and largely accounts for the work's popularity; but for any reader of an age after Latin ceased to be the vernacular or even the literary language, prodigious effort has been required to plod through Martianus' tortuous and neologistic bombast. The setting portions of *The Marriage* constitute some of the most difficult writing in the entire range of Latin literature.[9]

[6] *The Dark Ages*, p. 26.

[7] Another overstatement is found in P. R. Cole, *A History of Educational Thought* (Oxford, 1931), p. 78: "...the author, it is said, of the most successful textbook ever written." Cole does not indicate his source. A more reasonable judgment is that of Moses Hadas, *A History of Latin Literature* (New York, 1952), p. 406: "The loss of Martianus Capella, Boethius, or Cassiodorus Senator would have dealt the intellectual life of the Middle Ages a very severe blow."

[8] See Raby, *Secular Latin Poetry*, I, 100-1. Willis, p. 19, finds no indication that Boethius read Martianus "beyond perhaps glancing cursorily at the general layout of the work."

[9] Willis, who is at present engaged in preparing the new Teubner Library edition of Martianus, finds him the most difficult author he has ever tackled.

Comprehensive summaries of the narrative of Books I and II are readily available elsewhere,[10] but for the reader's convenience a few details will be offered here. Martianus avers at the outset of the story that it was told to him by Satire during long winter nights by flickering lamplight.[11] The story then begins: Mercury, after some unsuccessful attempts to secure a suitable wife, consults Apollo, who advises him to marry Philology, an astonishingly erudite young lady. The suggestion meets with the approval of both parties, and Philology, after considerable preparation and instruction, is wafted to the upper heavens, where her marriage is to take place before a "Senate" consisting of gods, demigods, and philosophers. The connection between the setting and the seven liberal arts becomes clear when an elderly but attractive lady named Grammar, one of the seven learned sisters, is introduced to present her discipline first to the assembled wedding guests. The seven sisters, personifications of the seven disciplines, have commonly been referred to as bridesmaids. They are bridesmaids only in the broadest sense of the word, however. Martianus calls them *feminae dotales* and, if we consider his fondness for legal vocabulary, the term should be translated as "ladies constituting a dowry." That is what they actually are: handmaids presented by Mercury to his bride. The marriage of Mercury and Philology has been taken, both early and late, to symbolize the union of eloquence and learning, the arts of the trivium and the quadrivium.[12]

The appropriateness of a marriage of eloquence and learning would

[10] The fullest summary, not only of the introductory setting, but of the setting portions of the later books, is found in Duckett, *Latin Writers*, pp. 224-29. See also Raby, *Secular Latin Poetry*, I, 102-4; Dill, pp. 412-13; Wessner, cols. 2005-6; and Cappuyns, "Capella," cols. 839-40. A running account of the work, with interesting comments, is to be found in the early but important study of Martianus by C. Böttger, "Über Martianus Capella und seine Satira," *Neue Jahrbücher für Philologie und Paedagogik*, Suppl. Vol. XIII (1847), pp. 594-602.

[11] 2 (Dick 4. 18).

[12] It has been so taken by Remigius in his commentary on Martianus (ed. Lutz, II, 175) and by John of Salisbury (*Metalogicon* 1. 1; 2. 3) and by a modern expert on classical and medieval allegory (Curtius, pp. 76-77). G. Nuchelmans has traced the union of *sapientia* and *eloquentia* from Cicero's statement in the preface of his *De inventione* to the close of the twelfth century ("Philologia et son mariage avec Mercure jusqu'à la fin du XIIe siècle," *Latomus*, XVI [1957], 84-107).

have been apparent to any Roman down to the fall of the Empire. Cicero's *De oratore* sets the goal of ambitions for a Roman gentleman.[13] While the philosopher stood at the pinnacle of the Greek world, in Cicero's eyes the orator was still more exalted because he combined consummate rhetorical skill with a mastery of the entire range of human knowledge. After the Empire fell, it was pointless for schools to maintain the pretense of preparing young rhetoricians for a career in politics. Nevertheless, during the Middle Ages rhetorical studies and classical literary models continued to engage the attention of students because they constituted the only academic curriculum familiar to the Roman world.[14]

To medieval scholars the close relationship between trivium and quadrivium studies was much clearer than it is to us. We readily appreciate the ties binding trivium subjects to each other and quadrivium subjects to each other; but medieval students also sensed a relationship between the trivium and the quadrivium. For medieval scholars, as for the early Pythagoreans, subjects like arithmetic and geometry seemed to have a universal application, and the rhetorical arts were found as useful in expounding quadrivium subjects as in literary subjects. A discipline like metrics seemed to belong as appropriately to grammar or rhetoric as to the mathematics of harmony, though in Martianus' story (326) Minerva decides that the subject belongs to music. Dialectic is called the sister of Geometry because the proofs of geometric propositions are, like dialectic, a form of logical reasoning. Varro is quoted in Gellius (18. 15. 1) as indicating the tie between metrics and geometry. The setting and trivium books in Martianus' work contain many references and allusions to quadrivium elements. Though present-day authors of general histories of the separate liberal disciplines are aware that Martianus had an important role in determining the medieval curriculum, they have not shown precisely what the nature of that role was.

Another interesting but neglected aspect of *The Marriage* is its form. It is a pedantic novel, a genre dear to readers of antiquity and

[13] Book I, esp. secs. 5-6, 12-14.

[14] See the chapter "Medieval Schools to *c.* 1300," by Margaret Deanesly, in *CMH*, V, 765-79.

the Middle Ages who wanted to have their encyclopedic learning capsulated with a sweet coating. Here too Martianus played a key role. Before his time authors who wished to purvey popular learning in narrative form adopted the symposium genre,[15] the prototype for which was Plato's *Symposium*. By the time the Romans began imitating this genre, it had deteriorated in the hands of Hellenistic polymaths. Aulus Gellius' *Noctes Atticae*—a potpourri of scraps which he gleaned from writers like Varro and Pliny and put into mouths of guests at his fictional banquet, usually without accurate attribution to his source[16]—set the fashion for later Latin writers.[17] Martianus was strongly influenced by Gellius, and his heavy debt, direct or indirect, can be traced through parallels of phraseology and vocabulary. Martianus, like his contemporary Macrobius, could have used a banquet as the setting for his encyclopedia of the arts, but he made a wise decision in seeking another setting. To put seven long disquisitions into the mouths of seven guests at a banquet would not have been so attractive to readers as to have them presented by seven supernaturally wise bridesmaids at a heavenly wedding ceremony. Martianus' book became much more popular in later ages than Macrobius' *Saturnalia*—the last

[15] The author of the article "Symposium Literature" in the *OCD* cites only one title on this subject. There is need for a work tracing this genre from its early use by Plato and Xenophon through its development into a didactic form in the works of Greek polymaths of a later age: Plutarch *Symposiacs*; Lucian *Symposium*; Athenaeus *Deipnosophistae*; and the Roman imitators Aulus Gellius and Macrobius (in his *Saturnalia*). Pseudo-Plutarch *De musica* also belongs to the symposium genre. The dialogue takes place on the second day of the Saturnalia and the speaker remarks (1131c-d) that on the previous day grammar was the subject of conversation. It is unlikely that the seven disciplines were the topics at this Saturnalia banquet. The *De musica* bears scant resemblance to a handbook on the subject.

[16] On the careless character of Gellius' compilations and his spurious attributions see the penetrating essay "The *Noctes Atticae* of Aulus Gellius" in Nettleship, *Lectures and Essays*, esp. pp. 252-58.

[17] A notable case was Macrobius' Saturnalia. Gunnar Lögdberg, *In Macrobii Saturnalia adnotationes* (Uppsala, 1936), has traced Macrobius' pillaging of Gellius and others, much of it verbatim. Martin Hertz, in his *editio major* of Gellius' work (Berlin, 1885), I, v-liv, discusses the use of Gellius by later writers, into the Renaissance; and Marrou, *Saint Augustin*, pp. 105-6, 121-22, remarks upon the importance of Gellius' erudition to the compilers of the late Empire.

of the Roman literary banquets—and with his innovation he established a new genre.[18]

Martianus' inspiration for his setting came from Varro's *Menippean Satires*[19] and from the Latin novelists who are influenced by Varro. Petronius' *Satyricon* and Apuleius' *The Golden Ass* are works which are rightly regarded as classics in the history of the novel. Both writers were brilliant artists who did not mar their stories in order to purvey learning; Petronius did the very opposite, cleverly satirizing the pedantry of the rhetoricians of his day. But Apuleius had another side which was naturally attractive to Martianus, a side that Martianus knew well. To his contemporaries, Apuleius was better known as a Platonist and a man of encyclopedic learning than as a novelist. He went about lecturing on all sorts of subjects. He was criticized in a later age for having indulged in the frivolities of fiction.[20] That Martianus was familiar with the minor works of Apuleius, writings on popular philosophy and erudition, can be traced in his vocabulary.[21] In considering the influence of Apuleius upon Martianus we must regard Apuleius in both of his literary roles, as a novelist and as a polyhistor, and in this light we must place him alongside Varro as a major source of inspiration. Thus Martianus' two prototypes were men of wit and learning, but Martianus was the first to combine these features of both in a Menippean fiction. His influence as a genre-writer and his relationship to his sources Varro, Gellius, and Apuleius are examined below, in the sections on Sources and Influence.

[18] Curtius, in "Jest and Earnest in Medieval Literature," pp. 417-20, sees Martianus as a representative of still another late-antique, early-medieval literary convention, the mingling of serious and comic styles.

[19] The author of the article "Symposium Literature" (see n. 15, above) sees a relation between the symposium and Menippean genres.

[20] Macrobius (*Commentary* 1. 2. 8) is surprised to find Apuleius engaging in such light sport. Marrou, *Saint Augustin*, p. 113, thinks that the lost works of Apuleius on the scholarly disciplines were known to Augustine, and the possibility exists that these works were also known to, and used by, Martianus.

[21] For many of the entries in Lewis and Short's *Latin Dictionary*, the only author cited besides Martianus is Apuleius.

PROSE STYLE

Any reader of Martianus, whether experienced in matters of style or not, will quickly conclude that Martianus wrote in a most extraordinary manner. His diction, abounding in neologisms, requires the constant use of lexicons; and his unwieldy sentences, loaded with the bombast and metaphors said to be characteristic of the "African school" of writers but carried by him to absurd extremes, caused at least one scholar to doubt that the author was seriously trying to communicate.[22] Martianus was himself aware of the barbarity[23] of his style, and on several occasions he expressed reservations about his abilities as a writer. Two passages might be quoted here as aptly describing his own style and indicating his misgivings—and, as providing, in their very mode of expression, justification for those misgivings:

memorans frigente vero
nil posse comere usum
vitioque dat poetae
infracta ferre certa
lascivia dans lepori
et paginam venustans
multo illitam colore

[221; Dick 81. 3-9]

disciplinas cyclicas
garrire agresti cruda finxit plasmate

[998; Dick 534. 1-2].

The theory of an African school of writers (*Africitas* and *tumor Africus*), first conceived by Renaissance scholars, has long since been repudiated, even by Karl Sittl, who was for a time its leading modern

[22] Teuffel, II, 448.

[23] Joseph Justus Scaliger, the outstanding Renaissance critic of Latin letters, called Martianus "Barbarus scriptor." See G.W. Robinson, "Joseph Scaliger's Estimates of Greek and Latin Authors," *Harvard Studies in Classical Philology*, XXIX (1918), 160; on p. 133 Robinson observes: "The recognized position of Joseph Scaliger as the greatest scholar of modern times—if not indeed of all time—gives a peculiar value to his estimates of the authors of classical antiquity."

proponent;[24] but speculations about African stylistic characteristics and their origins will continue. The neoteric propensity that Martianus was influential in transmitting to the Middle Ages has been traced to Apuleius, Gellius, Fronto, and Tertullian.[25] All but Gellius, whose birthplace is unknown, were Africans, and Sittl thought that Gellius too was an African. All wrote Greek with fluency, and Raby suggests that their pomposity may have been derived from Asiatic rhetoric.[26] When one notes how many unusual words of Martianus are cited in the lexicons as occurring only in African writers, one can readily appreciate why scholars continue to be intrigued with identifying the "African" qualities of Martianus.

Let us first consider Martianus' unusual diction without regard to its "African" character. Lexicographers have not yet really come to grips with Martianus. The manuscripts are in a very corrupt condition, and orthographical variants and textual cruces constitute a formidable problem for a lexicographer with standards for accuracy. Much scanning of manuscripts of texts, glosses, and as yet unedited commentaries remains to be done in order to check the work of previous lexicographers, marred as it is by omissions and dubious readings and interpretations. The day may never come, even in the computer age, when we will be in a position to label a word in Martianus' work definitively as *hapax legomenon.*

The original compilers and subsequent editors of Lewis and Short's *Latin Dictionary*[27] made a serious effort to include Martianus' vocabulary, but their coverage—in citing his peculiar words and their occurrences—is far from complete, and numerous incorrect interpretations

[24] Sittl supported the "African" theory in *Die lokalen Verschiedenheiten der lateinischen Sprache mit besonderer Berücksichtigung des afrikanischen Lateins* (1882). His retraction came ten years later. Einar Löfstedt, *Late Latin*, p. 42, regards the theory as merely of historical interest, in that it points to the consequences of faulty methodology.

[25] See Löfstedt, p. 1.

[26] *Secular Latin Poetry*, I, 21-22. On earlier writers, (esp. Apuleius), who influenced Martianus' style see Wessner, cols. 2006-7. See also Richard Newald, *Nachleben des antiken Geistes im Abendland bis zum Beginn des Humanismus* (Tübingen, 1960), pp. 145-46.

[27] Unabridged, the first edition published in 1879; reprinted frequently by the Clarendon Press, Oxford.

and brandings of words are given. Alexander Souter's *Glossary of Later Latin to 600 A.D.*[28] corrects many of the meanings in Lewis and Short and includes many words omitted by them, but it also omits words and has some doubtful or incorrect interpretations. The most thorough study of Martianus' vocabulary undertaken to date is found in the *Thesaurus Linguae Latinae*,[29] but it too has omissions. It is in progress and is only about half completed, and it covers authors and documents to around 600.

It is becoming apparent that Martianus, more than any other writer, was responsible for the fondness for unusual vocabulary evident in the writings of the Carolingian masters, and that he was one of the most pillaged of late Latin authors in the Middle Ages. His orthography, coinages, and usages will have to be checked in the commentaries and glosses on his work in later ages down to the Renaissance, and many more lexicons will have to be consulted.[30] For the purposes of the present study the works of Lewis and Short, Souter, and the *Thesaurus Linguae Latinae* will provide an ample basis for sampling.

One can readily appreciate Martianus' importance in the history of the Latin language by examining lists of his *hapax legomena* and rare usages and reflecting upon the enormity of those lists, in both the number and the nature of the examples. The freest use of *hapax legomena*, as might be expected, occurs in the first two books, devoted to the author's fantastic description of the bridal preparations and setting. A doctoral dissertation by Friedrich May[31] includes a thorough examination of Martianus' vocabulary in those books. May's lists and the list for Books VI-IX in Appendix B of this volume will indicate to anyone versed in Latin linguistics that it would be hard to find a Latin author with a more unusual vocabulary. (May's lists for Books I and

[28] Oxford, 1949.

[29] Leipzig, 1900-. Referred to in the notes as *TLL*.

[30] A comprehensive list of lexicons will be found in Karl Strecker, *Introduction to Medieval Latin*, pp. 38-45. See also Leonardi, "Raterio e Marziano Capella," *IMU*, II (1959), 93. To these must now be added the *Revised Medieval Latin Word-List*, ed. R. E. Latham. Also important for Martianus' neologisms is Dick, *Die Wortformen bei Martianus Capella*. Dick did the excerpting of Martianus for the *TLL*.

[31] *De sermone Martiani Capellae (ex libris I et II) quaestiones selectae*; see esp. Part III, "De vocabulorum copia," pp. 81-94.

II, for which Apuleius was Martianus' chief model and inspiration, show more clearly than the list for Books VI-IX Martianus' heavy debt to Apuleius.) Because of the influence Martianus had upon the Middle Ages, he deserves a place among the Latin writers who profoundly affected the course of linguistic development—Plautus, Cicero, Petronius, Fronto, and Apuleius.

Two types of neologisms are conspicuous in the quadrivium books of Martianus: (1) bold compounds, mostly from recognizable bases—a characteristic of the African writers—as seen in such words as *cerritulus*, *cuncticinus*, *hiatimembris*, *latrocinaliter*, *marcidulus*, and *perendinatio*; and (2) technical or scholarly words, such as *acronychus*, *apocatasticus*, *egersimon*, *interrivatio*, *Latmiadeus*, *metaliter*, *submedius*, and *trigarium*, many of this type being Latinized forms of Greek words. Fulgentius and the author of the *Hisperica famina* were especially attracted by Martianus' coinages.

The problem of reconciling Martianus' known popularity with the abstruseness of his sometimes technical, sometimes flamboyant, usages remains to be investigated. The more abstruse we regard him, the more difficult it becomes to explain his popularity. He appears to have come into a sudden vogue in the third generation of the Carolingian Age. We can see from the commentaries recently published and from the abundant glosses in ninth-century Carolingian hands that Martianus sparked the imagination and scholarly instincts of the learned commentators teaching in the schools founded by Charlemagne. These commentators explicated words and passages that would otherwise be unclear to us. Obviously their glosses did not originate with them, however. If we had a manuscript tradition for the earlier centuries, we would be able to trace the use of Martianus' textbook and the glossing of his text by schoolmasters during the early Middle Ages. The apparent suddenness of his vogue in the ninth century may be partly explained by the large number of manuscripts extant from that period.

The abundance of the glosses indicates that students had great difficulty in comprehending Martianus' text; it is also evident that the schoolmasters could provide an explanation, even if they had to resort to the standard practice of suggesting several alternatives without noting a preference. They delight in displaying both the range and meticulousness of their scholarship, coining new words and terminations to develop finer distinctions of meaning. Tendencies that seem reckless in Martianus go wild in the Carolingian commentators.[32] Remigius, for example, in commenting upon *aquilonalia* (Martianus 838) says: *id est aquilonaria, aquilonium, vel aquilonare signum* (ed. Lutz, II, 265); and upon *epistularis* (Martianus 896): *id est cancellaria, legataria, scriptrix . . . exceptrix* (ed. Lutz, II, 302). Martianus serves as an authority for the specialized and learned vocabulary of the Middle Ages, as Plautus and Petronius do for the colloquial dialects of classical times.

Martianus' bizarre style of composition inevitably suggests comparison with the rococo style of Apuleius' *Golden Ass*. Some critics have felt that Martianus was consciously emulating the style of his fellow-townsman. If this is true, he was far too successful. The straining for grand effects through ornate and recondite vocabulary, the conceits and soaring flights of imagination that characterize Apuleius' fiction style, an extraordinary one for his day, are restrained in comparison with the turgid and flamboyant style of Martianus.

Like his model, Martianus has two styles.[33] When he is purveying stock handbook materials he uses a fairly prosaic, intelligible style. In the setting portions[34] and when he attempts spectacular demonstrations of complex points, he uses a florid style, seldom matched by any Latin writer. The occasional departures from traditional handbook materials are introduced to impress his readers with the occult, primeval wisdom and divine intelligence possessed by his bridesmaids. Martianus himself makes the intent of these interpolations clear by calling our attention to the fact that a bridesmaid is speaking. The interpolations tricked both medieval readers, who introduced corruptions

[32] Raby, *Secular Latin Poetry*, I, 22, sees Martianus' role as crucial in inspiring later generations of glossators. Raby's perceptive comments are remarkable—considering that they were written before the Carolingian commentaries on Martianus had been published. H. Liebeschütz, in *The Cambridge History of Later Greek and Early Medieval Philosophy*, pp. 576-77, points out that the Carolingian scholars were fond of compiling polyglot vocabularies of the sacred languages and found in Martianus' abundant Greek terminology a fruitful field of study

[33] Apuleius uses a distinctly plainer style in his learned writings than in his fiction.

[34] Books I and II and the beginning and end of each subsequent book. The closing setting of Book VIII is missing in the manuscripts.

into the text in an effort to apprehend his meaning, and modern scholars, who, mistakenly regarding the passages as containing genuine information on the disciplines,[35] have undertaken to emend the corruptions in the text. Although Martianus' ornate style is one of the most difficult in all of Latin literature, his handbook style is not difficult to follow if one is already familiar with the precepts of his disciplines from having read them in other classical authors. When Martianus is unclear in his exposition because he himself has not comprehended the matter, the informed reader can repair the deficiencies of Martianus' writing or comprehension. The commentators sometimes provide substantial assistance, but much more frequently they are mistaken in their interpretation or explanation.

Martianus' obscure and florid style repeatedly reveals his inadequacies as a writer. His embarrassment is most evident in the purely descriptive passages. His depiction of the sensuous charms of the seven learned maidens is sometimes comical in its ineptness. When he describes the preparations and setting for the celestial wedding he resorts to circumlocutions, abstractions, and extravagant adjectives, because his powers of expression are far too feeble to achieve the sublimity he desires.

Near the opening of the story (11; Dick 10. 11-22), for example, Mercury and Virtue are looking for Apollo. They come to his "Delphic retreat" (*Cirrhaeos recessus*) and find "talking caves of hallowed hollow" (*sacrati specus loquacia antra*); they are "pressed by something or other of the ages, standing about in ranks" (*circumstabat in ordinem quicquid imminet saeculorum*); these are the "Fortunes of cities and nations, of all kings and all humanity" (*Fortunae urbium nationumque, omnium regum ac totius populi*). Some Fortunes have completed their course and are in flight; others are standing before them; still others are approaching. "The intervening distance causes some to appear indistinct, so that an incredible exhalation of smoky haze envelops them" (*nonnullis eminus vanescebat disparata prolixitas, ut velut fumidae caligationis incredibilis haberetur aura*). The final quotation might serve appropriately on the flyleaf of Martianus' book.

As the last bridesmaid, Harmony, is being readied to present her

[35] As I pointed out above (n. 22), however, at least one modern scholar has expressed doubts about Martianus' seriousness.

discipline, there is an agitated debate about the presentation of other learned maidens. Phronesis [Prudence], mother of Philology, has brought along seven other maidens, armed with lavish gifts, to be added to the dowry. The damsels "have been instructed in the more mystic and holy secrets of the maiden" (*in penetralibus quoque virginis et sanctioribus alumnatae*). It would be shameful to pass over them. The decision is left to Jupiter. The maidens, who turn out to be the seven prophetic arts, are introduced and the arts of the first three are explained. Next comes "the triplicity of the always supplicating sisterhood" (trigarium semper supplicantis germanitatis). After eight books we are conditioned to Martianus' abstractions, but he exasperates us here by neglecting to tell us who these three sisters are. The wedding guests (and Martianus' readers) are weary, Luna is unimpressed by the girls' credentials, and, "reminded of the day's twin division," she speaks up in the characteristically pompous and abstract diction of Martianus' celestial dignitaries:

I confess, I would like to learn the doctrines of such celebrated erudition if there is authority to defer the examination of the maidens, especially since a postponement until the day after tomorrow is a reasonable delay; thus might tedium and a distaste, occasioned by the fatigue of learning and the effort of concentration, not preclude all discourses of a subtle nature; and the keenness required for seeking and approving knowledge might not be blunted and turned into an aversion by complexity and prolixity.

A translation can scarcely preserve the flavor of the original.[36]

Martianus has been sprung on unsuspecting students of medieval Latin by mischievous teachers as a corrective for the notion that it is easy to read the debased writing of the post-classical period. To inflict a heavy dosage of Martianus even upon a promising graduate student, however, could cause him to abandon thoughts of a career in classical philology.

[36] 897 (Dick 475. 8-16): *et fateor vellem, si qua examinationem virginum prorogaret auctoritas, ipsa quoque tam praecluis eruditionis asserta cognoscere, praesertinque cum perendinatio rationabiliter expectatur, ne lassata cognoscentis curae fatigatione fastidia omnem doctae intimationis excursum gravatae laboribus intentionis excludant, et illa expetendae cognitionis adprobandaeque subtilitas in odium noscendorum obtusa multiplici prolixitate vertatur.*

One example of Martianus' awkwardness in delineating characters will suffice. Satire, the teller of his tale, is referred to only three times in all—once, at the opening, by the mere mention of her name. At the second reference (576; Dick 287. 608), though she chides Martianus for his dullness and stupidity, she is described as "charming" (*lepidula*) and "mirthful" (*iocabunda*). The third time (806-9; Dick 425. 19 - 427. 12), she again harshly rebukes Martianus, who notes, however, that "on other occasions she is charming" (807; Dick 426. 11). We never do see the charming and mirthful side of Satire.

Martianus' bridesmaids personify their disciplines and have absolute knowledge of their subjects. Euclid, Archimedes, and Ptolemy were among their prize pupils. Philology represents the mastery of all knowledge. These pontificating creatures belong to the literature of the Middle Ages and not to classical mythology. Geometry has made a flight to earth and has trodden upon all parts of its surface before giving an account of her discipline. She has also calculated the distance, even to fathoms and inches, between earth and the celestial sphere. Astronomy has flown through all the heavens and is familiar with their sectors; but, unfortunately for medieval readers, she will limit her discourse to phenomena observable to inhabitants of the northern hemisphere. She will divulge other secrets of the ages, however, observations accumulated by Egyptian priests and laid away in their sanctums, where she spent forty thousand years in seclusion. When Arithmetic appears on the scene, countless rays emanate from her brow in a miraculous proliferation and her calculating fingers whir like a hummingbird's wings. Harmony of course presents the Greek conception of her discipline, harmonic theory, the mathematics of music, but she has also introduced all forms of instrumental music upon earth. The symphony that marks her entrance is a harmony of the planetary and celestial spheres. In depicting her and her musical accompaniment (909-10) Martianus puts his powers of description to the supreme test; he concludes with the remark that the symphony is impossible to describe.

At the opening of Book II, Philology, by now appointed to be the bride of Mercury, is preoccupied with thoughts of her forthcoming marriage. She has always had a passion for Mercury though she has caught only fleeting glances of him—as he ran about the palaestra after an oil rubdown, while she was gathering flowers outside.[37] Philology's apprehensions (*anxia dubitabat*) about the propritiousness of the marriage are reminiscent of Dido's wavering about a union with Aeneas. Philology, too, will seek wisdom in omens—in her case, in the ancient lore of numerology. She rapidly calculates the numbers represented by her name and Mercury's, not one of the common names of Mercury but the divine one given to him by Jupiter (Martianus alludes to the name but does not state it).

She took from each end of Mercury's name the bounding element that is both the first and the perfect terminus of number. Next came that number which is worshiped as Lord in all temples, for its cubic solidity. In the next position she took a letter which the Samian sage regarded as representing the dual ambiguity of mortal fate. Accordingly, the number 1218 flashed forth. (*Ex quo finalem utrimque sumit, quae numeri primum perfectumque terminum claudit; dehinc illud quod in fanis omnibus soliditate cybica dominus adoratur. Litteram quoque, quam bivium mortalitatis asserere prudens Samius aestimavit, in locum proximum sumit, ac sic mille ducenti decem et octo numeri refulserunt* [102; Dick 43. 13-19].)

This passage was an enigma to the Carolingian commentators, who strained to discover Mercury's "divine name."[38] The first writer to explain it was Hugo Grotius,[39] the prodigy who, at the age of sixteen, published the most important edition of Martianus (Leiden, 1599) before the nineteenth century; if Grotius had remained in classical philology, he might have become a scholar to rank with Scaliger.

The number that is "both the first and the perfect terminus of number" is 9, which completes the first Pythagorean series.[40] The

[37] Richard Johnson, the contributor of the trivium section in this volume, and my collaborator in the forthcoming translation of *The Marriage*, understands the flowers as symbolizing the rudiments of grammar and literary study.

[38] Remigius *ad loc.* (ed. Lutz, I, 146-47) and John Scot Eriugena *ad loc.* (ed. Lutz, pp. 56-57) suggest ΧΥΡΡΙΗ as Mercury's "divine name." The numerical values of the Greek letters, added together, give 1218; and the resemblance of the name to *Kyrie* undoubtedly recommended it to Christian readers.

[39] See his notes on this passage (keyed to the Latin text on p. 25 of his edition).

[40] In discussing the attributes of each number of the Pythagorean decad, Martianus later points out (741) that the number 9 marks the end of the first series; 10 also marks the end of the first series (742), but it belongs to the second series as well.

number that is said to be worshiped "for its cubic solidity" is 800.[41]

The Greek letter which, according to the "Samian sage" (Pythagoras), represents the "dual ambiguity of mortal fate" is the crossroads letter (Y). Putting together the Greek letters which represent the numerals described Grotius got Thouth, or Thoth, the Egyptian name of Mercury.[42] The sum of the numbers in Philology's calculation is 1218. Similarly, taking the numerical values assigned to the Greek letters in her own name *Philologia*, she gets a sum of 724. "Reducing both numbers by the rule of nine,"[43] she gets a remainder of 3 for Mercury's number and 4 for her own. These two numbers are among the most revered in the Pythagorean decad and, in all their associations and attributions, portend harmony (*jugitas*).[44]

While the oracular tones and technical jargon of the bridesmaids must have impressed Martianus' contemporaries, these devices undoubtedly also concealed Martianus' deficient understanding of the subject. Such would have been the case in his explanation of Eratosthenes' method of measuring the circumference of the globe (596-98), a simple geometric procedure that no Latin writer, ancient or medieval, understood.[45] We are surprised not so much at the ineptness of Martianus' explanation as at his boldness; he was the first Latin writer known to have attempted to explain Eratosthenes' method. Later, Astronomy interrupts her prosaic presentation of handbook materials (858) to address her readers in the first person. She "reminds" us that her sister Geometry had previously demonstrated (596) the determination of the earth's circumference at 406,010 stadia. (Geometry's figure

[41] Actually 8 represents the cube in Pythagorean numerology, as Martianus (740) and Macrobius (*Commentary* I. 6. 3) point out. See also *Nicomachus of Gerasa Introduction to Arithmetic*, tr. M. L. D'Ooge, p. 106. But Martianus is thinking of 800 as the product of 8 and 100, the latter being the square of 10.

[42] $\Theta = 9$; $\Omega = 800$; $Y = 400$; $\Theta = 9$. On Mercury's identity with Thoth and on his discovery of the arts in Egypt see Cora E. Lutz, "Remigius' Ideas on the Origin of the Seven Liberal Arts," pp. 34-35.

[43] This rule was known to classical mathematicians. When any number is divided by 9, the remainder is the same as that left when the sum of the digits in the original number is divided by 9. For example: 8573 divided by 9 leaves a remainder of 5; likewise, the sum of 8, 5, 7, and 3—that is, 23—when divided by 9 leaves a remainder of 5.

[44] 103-8 (Dick 43. 19 - 46. 7).

[45] See below, pp. 134-35.

was actually 252,000 stadia.) Astronomy then proceeds (859-61) to show, by comparing measurements of shadows during eclipses, the relative size of the earth and the moon—she is comparing the earth's circumference and the moon's diameter, however—and from this basis to measure the orbits of all planets, an undertaking not even attempted by Ptolemy. Macrobius makes a similar departure from his handbook materials, expressing contempt for the efforts of Eratosthenes and Posidonius to ascertain the dimensions of the sun and attributing the method revealed by himself to the "canny Egyptians" (*Aegyptiorum sollertiam*).[46]

We may find the hauteur and jargon of Martianus' female wizards repellent, but medieval readers were impressed. Moreover, they were also titillated by his spicy descriptions of their charms and by other fillips which he added to the setting. To judge by Remigius' appreciation of such details, Martianus' innocuous lasciviousness contributed in some small measure to the popularity of his book.[47]

The conventional subject matter of the disciplines is also delivered by the bridesmaids in the form of discourses, but the reader quickly forgets the setting as Martianus addresses himself to the fundamentals. Were it not for occasional ornateness of diction or phrase, the presentation might resemble a conventional Latin handbook on the subject. Martianus is systematic in his arrangement of topics and gives each of them proportionate treatment. Terms are defined in a professional manner, and after classifications are set up, the terms are given a second and fuller treatment.[48] Martianus' handling of the disciplines

[46] *Commentary* 1. 20. 9-31. Elsewhere Macrobius undertakes to explain the discrepancy between Plato's and Cicero's orders of the planets and ascribes one system to the Egyptians, the other to the Chaldeans. His account of the orbits of Mercury and Venus is so ambiguous that it has been mistakenly assumed by historians of science to refer to the system of Heraclides of Pontus. See *Macrobius Commentary*, tr. Stahl, pp. 162-64; 249-50.

[47] Martianus 726 (Dick 364. 15-20), Remigius *ad loc.* (ed. Lutz, II, 176); Martianus 727 (Dick 364. 21 - 365. 4), Remigius *ad loc.* (ed. Lutz, II, 176); Martianus 889 (Dick 471. 10-13), Remigius *ad loc.* (ed. Lutz, II, 296-97).

[48] These and other standard practices of ancient handbook compilers were pointed out for the first time by M. Fuhrmann, *Das systematische Lehrbuch.* Fuhrmann regards these handbooks as constituting a distinct literary genre. The chief defect of his monograph is that it deals almost exclusively with rhetorical

must have been satisfactory to his readers, and the main reason for the popularity of his book.

VERSE STYLE

Critics who have had harsh things to say about Martianus' prose have spoken appreciatively of his poetry. Themselves schooled in classical models, they have applied classical standards as criteria. There is a sharp contrast between Martianus' rococo propensities as a prose stylist and his usually correct handling of meter and high regard for classical models in poetry. His poetic diction, however, is as bizarre as his prose diction; and his poetry is even harder to translate than his prose, largely because of the diction. The textual cruces resulting from the difficulties scribes had in understanding his meaning add to the problems of the translator. Although the technical aspects of Martianus' meter have been thoroughly studied,[49] much remains to be considered about the poetic qualities of a writer who should be regarded as a significant figure in the history of Latin poetry. As a representative of the classical Menippean *prosimetrum* (mixed prose-verse) genre, he stands with Varro, Petronius, and Seneca.[50] In the Middle Ages, he and Boethius were the two principal models for the chantefable genre.

Martianus' competence and conservatism as a technician and his skillful handling of many meters were the main reasons offered by Dick and Wessner for placing him in an earlier century than the one to which he is now assigned. He uses fifteen different meters in all,[51]

---

manuals; the author is unaware that quadrivium handbooks conform to the type as well as trivium handbooks.

[49] F. O. Stange discusses each metrical type and shows in tabular form how Martianus imitated his predecessors; A. Sundermeyer, *De re metrica et rhythmica Martiani Capellae*, reviews the work of Stange in the first forty pages, then continues an analysis of the metrical forms. Dag Norberg, *Introduction à l'Étude de la versification latine médiévale* (Stockholm, 1958), pp. 71, 73-74, 78, discusses Martianus' use of caesura and some of the rarer verse forms.

[50] In the opinion of J. W. Duff (*Roman Satire* [Berkeley, 1936], p. 104). There is too little verse in Apuleius' *Golden Ass* for it to be considered an example of Menippean satire.

[51] These are classified, with references, by Schanz, Vol. IV, pt. 2, pp. 169-70. Wessner, col. 2007, gives a count of the number of lines found in each type.

and in the wide variety of his meters is classed by Stange with Ausonius, Prudentius, and Boethius.[52] Dactylic hexameters, elegiac distichs, and iambic senarii predominate in *The Marriage*. One poem, consisting of twenty-eight elegiac pentameters, would be of particular interest to students of metrics. A few false vowel quantities in Martianus' poetry evidence the trend which began in late antiquity toward the substitution of accent for quantity as the determining rhythmic element.[53]

Stange and Dick have collected instances of Martianus' borrowings from classical poets. Any new investigator of course will be able to add to their collections.[54] Doubts rise here, as with any similar gathering of instances, about whether some of the parallels cited represent actual borrowing: *insomni ... cura* (Lucan *Pharsalia* 2. 239) and *insomnes ... curas* (Martianus 579) raise understandable doubts; but when *saxificam ... Medusam* occurs in the next line (cf. Lucan 9. 670: *saxificam ... Medusam*), one is more favorably inclined to the possibility of Lucan's influence. Some other isolated parallels suggested by Stange and Dick are highly dubious. As might be expected, Martianus' indebtedness to Vergil, from all three of his major poems, is far greater than to any other poet. His borrowings follow the pattern of the borrowings of poets in the intervening centuries and of Vergil's borrowings from Lucretius: two or three words in juxtaposition or proximity are borrowed, but their order is often juggled and their cases changed. After those from Vergil the clearest instances of poetic borrowing in the quadrivium books are from Lucan and Ovid. There is one likely instance from Cicero's *Aratea*.[55]

[52] Stange, p. 3.

[53] E.g., *axiŏma* (327); *flăgitaret* (908). On the trend in Latin literature see the remarks of A. C. Clark in the preface of his *Fontes Prosae Numerosae* (Oxford, 1909).

[54] Morelli, pp. 252-55, lists many omissions of Stange, but the greater number of his additional instances seem dubious to me.

[55] *Aratea* 644: *luce Bootes* (Martianus 808: *luce Bootes*).

Sources

PRESENT-DAY SCRUPLES regarding the use and acknowledgment of sources are not applicable to the loose practices of Latin compilers. *Compilatio* means "plundering" or "pillaging."[1] The compiler was a poseur, an encyclopedist pretending to a mastery of his subjects while skimming the surface. Concealment of his sources was in keeping with the character of his learning. To have indicated the true nature of his borrowing would have shattered his image as a great authority. If predecessor A had cited B as his source, the compiler never cited A as his actual source and felt honorable in citing B as his own source. When fifth-century writers like Sidonius and Martianus cited as authorities Orpheus, Pythagoras, Fabius Cunctator, Cato, Thales, Democritus, Eratosthenes, Archimedes, and Ptolemy, they were following a practice that had begun with the early compilers of the Republic. Names of authorities came tagged to the bits of information with which they were identified as these *recepta* were passed down through generations of polyhistors. Pliny the Elder, in the Preface to his *Natural History*, expressed shock at finding that recent authors whom he regarded as most trustworthy had copied verbatim from their sources without naming them.[2] This he felt was tantamount to theft. Compared with other Latin authors Pliny was scrupulous about acknowledging his indebtedness, but he listed Egyptian, Babylonian, Persian, Etruscan, and Carthaginian authorities among his primary sources. Aulus Gellius, rather than Pliny, set the fashion for later Latin com-

[1] It would be interesting to trace the Romans' own attitudes about the ethics of compilation with concealment of sources. Isidore (*Etymologiae* 10. 44) informs us that the epithet *compilator* was given to Vergil by detractors who criticized him for adapting Homer's verses to his own use. Vergil's reply, as reported by Isidore, was that it took a mighty man to wrench the club from Hercules' hand. Cf. Suetonius *Vita Vergili* 46; Macrobius *Saturnalia* 5. 3. 17.

[2] Pliny *Natural History* 22.

pilers,[3] culling his excerpts at random and freely misrepresenting his sources.

Of the numerous sources Martianus drew upon, the most important for the narrative setting of *The Marriage* was Apuleius.[4] A lengthy discussion of Apuleius' influence upon Martianus would not be appropriate in this study, but a few remarks are in order since Martianus uses the allegorical setting to open and close each of the quadrivium books.[5] His main inspiration for the setting was the Cupid and Psyche episode occupying the middle books (4. 28 – 6. 24) of *The Golden Ass*. That episode is an old wives' tale (*anilibus fabulis*), told by a silly and tipsy old woman (*delira et temulenta anicula*) to a captive maiden; Martianus' story is an old man's tale (*senilem fabulam*), told by a mirthful and witty (*lepidula et iocabunda*) Satire to a silly old man (Martianus), captive in his retirement quarters during the long nights of winter. As Martianus' tale unfolds we are impressed with the many Apuleian reminiscences, particularly at the end, where the heavenly banquet celebrating the marriage of Philology and Mercury calls to mind the marriage feast of Cupid and Psyche. Both females are granted immortality. And it is from Apuleius that Martianus gets his penchant for introducing allegorical figures and personifications of the virtues. Such figures as Aurora, Cupido, Fides, Gratiae, Portunus, Psyche, and Voluptas, are common to both works.

Varro, who appears to have been the principal ultimate source of Martianus' encyclopedic learning, is a shadowy figure in Martianus' background.[6] The two works of Varro that seem to have had the

[3] Max Manitius, *Geschichte der lateinischen Literatur des Mittelalters*, I, 4, gives this key position to Gellius, pointing to his influence upon Macrobius and Martianus.

[4] Monceaux, pp. 453-54, discusses Martianus' indebtedness to Apuleius. See also Courcelle, *Lettres grecques*, p. 201; and p. 84 below.

[5] Although the closing scene for Book VIII is missing from the manuscripts, there is no reason to doubt that it conformed to the other books.

[6] *Ibid*., p. 199, relying upon the investigations of Eyssenhardt in his 1866 edition (xxxi-lviii), says of Martianus: "Sa dette envers Varron, même s'il ne l'a pas connu directement, est immense"; Courcelle also commends Wessner for his cautious attitude about Martianus' borrowings (Wessner, cols. 2007-12) and agrees with Wessner that Martianus was familiar with Varro's doctrines mainly through intermediaries.

greatest influence on Martianus have disappeared, and attempts at reconstructing them have produced scant results. Some six hundred lines survive from Varro's *Menippean Satires*, which originally numbered one hundred and fifty books.[7] The fragments of his *Nine Books of the Disciplines* have been painstakingly gathered from the works of later compilers and encyclopedic writers like Pliny, Gellius, and Augustine.[8] It is usually a moot question, in any single case, whether we are confronted with a direct or an indirect use of Varro as a source. Verbatim correspondences between phrases in Pliny, Gellius, Cassiodorus, and Martianus do not necessarily indicate that all of them used Varro's original work. Even less assurance is given by the correspondences between data in Book II of Pliny and in the closing sections of Martianus' eighth book. Varro's work was undoubtedly ransacked by compilers shortly after it was put into circulation. His learning filled contemporaries, even Cicero, with awe,[9] and a work of his, bringing the arts of the Greek *enkyklopaideia* into Latin form, must have aroused immediate interest. It has been conjectured that the work was quickly put into digest form. Eyssenhardt felt, after much analysis of the Varronian correspondences in Martianus, that Martianus did not consult Varro's original work.[10]

Despite the loss of both Varronian sources for Martianus, we have

[7] The fragments of the satires have been collected in *Petronii Saturae et libri Priapeorum: Adiectae sunt Varronis et Senecae Saturae similesque reliquiae*, ed. F. Buecheler and W. Heraeus (Berlin, 1912). J. W. Duff, *Roman Satire*, pp. 84-91; and Theodor Mommsen, *The History of Rome*, rev. ed. (New York, 1895), V, 486-92, offer a summary and discussion of the contents of the fragments. Martianus probably derived the name of the narrator of his story, *Satura*, from the title of Varro's work; and the title of one of Varro's satires, ὄνος λύρας [an ass listening to a lyre], taken from a Greek proverb, is coupled by Martianus (807) with another Greek proverb to produce a quotation of four words, the longest Greek passage in his book.

[8] Friedrich Ritschl, "De M. Terentii Varronis Disciplinarum libris commentarius," in his *Kleine philologische Schriften* (*Opuscula philologica*), III, 352-402, attempts a reconstruction of Varro's work.

[9] Duff, pp. 85-86, cites the instances in which Cicero shows his deep respect for Varro's learning.

[10] See n. 6 above. A recent book on Varro by Francesco della Corte, *Varrone: Il terzo gran lume romano*, contains a chapter on *The Nine Books of the Disciplines*, pp. 237-54.

a remarkably clear picture of the antecedents for the bulk of three of the quadrivium books, owing to the fortunate accident that five of the major sources of those books are extant. For only one of the quadrivium books (VIII) are the sources left largely to conjecture; this book is thought by most *Quellenforscher* to have been derived ultimately from Book VI (*De astrologia*) of Varro's *Nine Books of the Disciplines*.

BOOK VI: ON GEOMETRY

Martianus' *De geometria* is misnamed. The bulk of the book turns out to be a geographical excursus, consisting of a list of place names with random comments interspersed, the whole being introduced, in accord with convention, by a section on the principles of mathematical geography. Geometry, the expositor of the book, admits that she has been digressing when she finally addresses herself to the precepts of her art. It seems highly unlikely, therefore, that Varro's *De geometria*, the fourth book of his *Nine Books of the Disciplines*, was a major source for Martianus' book.

Ritschl, relying heavily upon a passage in Cassiodorus' *Institutiones*,[11] thought that Varro's book dealt with both the theoretical and the practical aspects of geometry and that the practical aspects he included were surveying and geography.[12] Boissier, while criticizing Ritschl's reconstruction of Varro's work as insubstantial, nevertheless supposed that the passage in Cassiodorus upon which Ritschl relied does divulge the contents of Varro's book.[13] Cassiodorus' statement must be handled with caution, however. He was obviously incapable of absorbing what he may have read on geometry in Boethius or any other writer and in the passage in question he was desperately seeking material to fill out a chapter on geometry. He succeeded in writing only two pages, the briefest treatment given by him to any discipline; only one short paragraph is devoted to geometry proper—the five

[11] *Institutiones* 2. 6. 1.

[12] Ritschl, pp. 385-90.

[13] G. Boissier, *Étude sur la vie et les ouvrages de M. T. Varron*, pp. 328, 333-34. For a bibliography of more recent studies on Varro's book see H. Dahlmann's article, "M. Terentius Varro," in Pauly-Wissowa, Suppl. Vol. VI, cols. 1255-59.

divisions of geometry, with definitions of each.[14] To introduce the subject, he quotes Varro on the derivation of the word *geometria* and states that the term was applied first to surveying boundaries, then to measuring years and months, astronomical distances, and the circumference of the globe.[15] Yet, considering Varro's general interest in etymologies and Cassiodorus' lack of interest in geometry, we cannot conclude from this passage that Varro's *De geometria* was primarily on surveying (*res gromatica*) and geography or that Cassiodorus detailed all the contents of the book.

A better informant about the contents of Varro's book is Aulus Gellius. In two chapters in his *Attic Nights* he relates in detail and with comprehension material that he found in Varro's *De geometria.* One chapter (16. 18) deals with two of Varro's divisions of geometry—optics and harmonic theory—and with the subdivision of harmony into rhythm, melody, and metrics. Another chapter (1. 20) gives several of Varro's definitions of geometric terms, using Greek words. Gellius points out that Euclid's definition of a line, which he quotes correctly, is more concise than Varro's—has anyone been as terse as Euclid in definitions?—and he suggests coining a Latin word *inlatabile* to express Euclid's ἀπλατές [breadthless]. We see from Gellius' brief quotations that Varro handled Euclid with skill and care and that he probably could have translated the *Elements* in its entirety, although this he surely did not do. There has been much study of the surviving Latin fragments of Euclidean geometry but no evidence to indicate that the whole work was translated into Latin before Arabic manuscripts of it became available to Latin readers.[16]

[14] Cassiodorus' divisions of geometry and his definitions correspond almost verbatim to those of Isidore *Etymologiae* 3. 11. 1 - 12. 1.

[15] In words that do not correspond to Cassiodorus', Isidore (3. 10) also introduces the subject of geometry by giving the derivation of the word and stating that the geometer's art was applied first to surveying lands and then to ascertaining the dimensions of the earth and the celestial sphere, and the distances (measured in stadia) between earth, moon, sun, and celestial sphere. Isidore, however, then continues for two pages to discuss geometric figures, whereas Cassiodorus, obviously embarrassed about discussing geometry, shifts to treating astronomy in the last paragraph of the chapter.

[16] George D. Goldat, *The Early Medieval Traditions of Euclid's* Elements, traces Latin versions and fragments of Euclid and provides a full bibliography.

What, then, did Varro's *De geometria* contain? Though Ritschl may have been correct in assuming that the book dealt with both the theoretical and the practical aspects of the subject, he was incorrect in supposing that geography was one of the practical aspects. Pliny's bulky excursus on the geography of the known world (*Natural History*, Books III-VI), which contains thousands of place names, is commonly believed to have been derived largely from Varro.[17] If this supposition is correct, Pliny must have gotten his data from other geographical writings of Varro and not from Varro's terse encyclopedia of nine disciplines in nine presumably normal-sized books. Gellius indicates that Varro handled Euclid's *Elements* in some detail, but there would have been room in one book for no more than a selection of Euclidean definitions, statements of propositions, and perhaps a sampling of proofs. That must have constituted Varro's geometry. Like Dante, who asks at the opening of his *De monarchia* why anyone should be interested in demonstrating once again a theorem of Euclid, Varro probably felt that geometry was too abstruse for Latin readers. It does appear from Cassiodorus' and Isidore's extracts that Varro included some fundamentals of mathematical geography in his book—the dimensions of the globe and the heavens and the distances between planets and the celestial sphere. If he gave the distances in stadia, the dimensions were based probably on musical intervals—beginning with an estimate of the moon's distance from the earth as 12,000 stadia—as is the case with the figures which Pliny[18] and Censorinus[19] attributed to Pythagoras but presumably derived from Varro. On the other hand, Ritschl was probably correct in assuming that Varro's book contained elements of the art of surveying. Varro was an eminently practical man, and to the Romans geometry was a subject to be studied for purposes of mensuration and reckoning.[20] The Varronian extracts in

---

To this may be added an important recent study by B. L. Ullman, "Geometry in the Mediaeval Quadrivium," in *Studi di bibliografia e di storia in onore di Tammaro de Marinis*, IV, 263-85.

[17] See J. O. Thomson, *History of Ancient Geography*, p. 218, on Varro's geographical writings.

[18] *Natural History* 2. 83-84.

[19] *De die natali* 13.

[20] Cicero *Tusculan Disputations* 1. 5.

Cassiodorus and Isidore point to the application of geometry to surveying, and the extant treatises of Roman *agrimensores*, or land-surveyors, either contain a modicum of geometry themselves or are bound with Euclidean fragments in the codices.[21]

Varro's *De geometria*, then, was not the main ultimate source for Book VI of *The Marriage*. However, the small part of Book VI which deals with Euclidean geometry[22]—a ten-and-a-half page compendium of Euclidean definitions, axioms, and propositions—may have been derived from some late, as yet unknown, digest of Varro's *De geometria*.[23] The sources of the geographical material in the book, on the other hand, can be more closely traced. The principles of mathematical geography and the list of place names with accompanying remarks were derived, probably through intermediaries, from Books II-VI of Pliny's *Natural History* and from Solinus' *Collectanea rerum memorabilium*. Mommsen's cogent argument that Solinus used for the *Collectanea* not Pliny's original work but a chorography based on it[24] suggests that Martianus too used a chorography based on Pliny's work.[25]

[21] See Ullman on the close tie between *ars geometriae* and *ars gromatica* in the medieval manuscripts. N. M. Bubnov (*Gerberti postea Silvestri II Papae, opera mathematica*, pp. 494-508) believes that the remains he found of writings of Roman *agrimensores* were from Varro.

[22] Martianus' introduction to geography and his list of place names occupy five times as much space (590-703) as his compendium on geometry (706-23).

[23] Martianus' definition of a line (*linea est, quam* γραμμήν *vocamus, sine latitudine longitudo*) differs from Varro's according to Gellius' transcription: *linea est longitudo quaedam sine latitudine et altitudine*. Gellius points out that Euclid's definition omits altitude.

[24] In his edition of Solinus' *Collectanea* (1895), pp. xvii-xix, Mommsen points out certain verbal correspondences between Solinus and Apuleius in passages that stem from Pliny. Mommsen shows that the passages differ from Pliny's phrasing and concludes that Solinus and Apuleius used the same chorography based on Pliny, rather than Pliny's original work. G. M. Columba, in two decades of publications entitled "La questione soliniana," collected in his *Ricerche storiche*, I, 171-352, discusses correspondences with the Roman geographer Mela (*c.* A.D. 41) and with Pliny in Solinus and develops the thesis that there was an earlier chorography, drawn upon by both Pliny and Mela, which influenced later traditions of Roman geography. Mommsen lists Mela-Solinus correspondences on p. 238.

[25] Eyssenhardt (*Martianus*, p. xxxi) feels, after Mommsen's demonstration, that Martianus did not use Pliny directly, and I would agree with him. Schanz, Vol. IV, pt. 2, p. 169, believes that Martianus used Pliny and Solinus directly.

Though it is probable that Pliny's books on cosmography (II) and geography (III-VI) were in turn drawn largely from two or more works of Varro,[26] there is no need to speculate about Martianus' relation to Varro in the field of geography. Once we pass Martianus' opening paragraphs, which give doxographical statements about the shape of the earth and proofs of its sphericity that could have been derived from Varro, we can compare nearly every statement he makes with some passage in Pliny or Solinus,[27] the majority of passages showing some verbal correspondence. Martianus' dependence upon a Pliny-derived chorography is greater in the earlier part of his geographical excursus than in the latter part. He may have had chorographies based on Pliny and Solinus before him throughout his compiling. Some new investigator, by carefully comparing the correspondences between Mela, the first extant Latin geographer, and Pliny, Solinus, and Martianus, may arrive at firmer conclusions about the derivation of these geographical compilations.

BOOK VII: ON ARITHMETIC

A rather similar situation obtains with regard to the sources for Martianus' *De arithmetica*. The immediate source or sources for Book VII are not known, but we have both ultimate sources to compare with nearly every statement Martianus makes on the subject of arithmetic. These sources are Euclid's *Elements* and Nicomachus of Gerasa's *Introduction to Arithmetic*. Eyssenhardt, in his 1866 edition of Martianus,[28] conjectured that Varro's fifth book of disciplines (*De arithmetica*) was the source for Martianus' discussion of the virtues and attributes of the numbers in the Pythagorean decad. That discussion occupies only a small portion of Book VII, however, and Eyssenhardt

[26] Which particular works Pliny used is a matter of debate. See Alfred Klotz, *Quaestiones Plinianae geographicae* (Berlin, 1906), p. 9.

[27] F. Lüdecke, *De Marciani Capellae libro sexto*, pp. 13-35, collects the Pliny-Solinus-Martianus correspondences and discusses Martianus' conflation of sources. Theodor Mommsen, (Solinus *Collectanea*, p. xxvi) commends Lüdecke on his accuracy, and on pp. 243-44 lists the Solinus-Martianus correspondences. Dick in the apparatus of his edition of Martianus lists nearly all the Pliny correspondences but omits nearly half of the Solinus correspondences.

[28] Pp. liii-lvi.

was unaware of Martianus' heavy dependence upon Nicomachean and Euclidean arithmetic for the rest of the book. Nicomachus did not compose his work until about A.D. 100, more than a century after Varro's lifetime. Of course the possibility exists that an exemplar of "Nicomachean" arithmetic had been in circulation a few centuries earlier and was worked into Varro's book; but Eyssenhardt was not aware of such considerations. Dick's list of Euclid and Nicomachus references in the apparatus of his 1925 edition is nearly complete. In the following year a thorough and comprehensive work on Nicomachus' *Introduction* appeared, tracing the traditions of his arithmetic from its early Pythagorean sources to the late Middle Ages.[29] It contains an epitome and discussion of Martianus' arithmetic as well as comparisons with the arithmetic of Cassiodorus and Isidore.[30]

Waszink, in his masterly edition of Calcidius' translation and commentary for Plato's *Timaeus*, lists several parallels between Calcidius and passages in Book VII of Martianus, but these are doctrinal resemblances and not verbal correspondences. With respect to compilers, only verbal correspondences constitute sufficient evidence of borrowing. Waszink conjectures that Calcidius lived around the close of the fourth century and would then have been a contemporary of Martianus. There is no evidence in Waszink's parallels that Martianus read Calcidius.[31]

[29] Nicomachus of Gerasa, *Introduction to Arithmetic*, tr. by M. L. D'Ooge, with studies in Greek arithmetic by F. E. Robbins and L. C. Karpinski. Also important in tracing traditions of Nicomachean arithmetic are earlier articles written by Robbins: "The Tradition of Greek Arithmology," *CP*, XVI (1921), 97-123; and "Posidonius and the Sources of Greek Arithmology," *CP*, XV (1920), 309-22. These important Nicomachean studies are not well known to Continental scholars, who, during the past century, have shown much greater interest in the popular science and philosophy of the ancient and medieval worlds than British and American scholars have.

[30] Nicomachus (tr. D'Ooge), pp. 138-42.

[31] Calcidius (ed. Waszink) 76. 4-5 (Martianus 105, 733); 85. 17, 18 and 86. 12-87. 1 (Martianus 738); 86. 8-11 (Martianus 739); 165. 16-19 (Martianus 736). In one case (86. 8) there is verbal correspondence. Calcidius: *sunt in capite septem meatus*, Martianus: *septem meatus habet in capite*. But it is clear from the list of seven vital organs two lines later that Calcidius was not Martianus' source for this discussion. Calcidius: *lingua pulmo cor lien hepar duo renes*; Martianus: *linguam cor pulmonem lienem iecur et duo renes*. Besides, it should be pointed out that in the

BOOK VIII: ON ASTRONOMY

The sources of Martianus' *De astronomia* are open to speculation. Every historian of science and philologist who, to my knowledge, has remarked upon this matter has conjectured that Varro was Martianus' source, either ultimate or direct. The evidence that they offer is of the bits and pieces variety, and even as an aggregate it is inconclusive; these fragments do not provide a reconstruction of Varro's *De astrologia*. Yet it must be admitted that Varro has no rival as a likely source of Martianus' astronomy. Tracing sources beyond Varro is risky, but *Quellenforscher* have found it challenging and intriguing. They have undertaken reconstructions of the cosmography of Posidonius, Varro's commonly presumed source. Posidonius was the grand figure of late Hellenistic popular science, and the extant handbooks of Geminus, Theon, and Cleomedes are in the Posidonian tradition.[32] Correspondences among them and between them and Latin writers on cosmography—Pliny, Censorinus, Macrobius, Calcidius, and Martianus—are striking. These similarities lead most scholars to suspect that Latin cosmographic traditions go back to Varro, and beyond him to Posidonius; but the arguments tend arbitrarily to rule out other possibilities.[33]

---

writings of compilers there is probably greater correspondence in passages dealing with associations of the numbers of the Pythagorean decad than in any other subject.

[32] A recently published doctoral dissertation, remarkable for its care as well as its length (579 pp.), presents a thorough study of the fragments of Posidonius, a critical review of previous reconstructions of Posidonius, and separate chapters on each of the scientific disciplines: Marie Laffranque, *Poseidonios d'Apamée* (Paris, 1964).

[33] Popular handbook traditions began to develop even before the Hellenistic Age and in the hands of Eudoxus and Eratosthenes took on very definite shapes. Some of the traditions originating with these two commanding figures may have bypassed Posidonius. It was fashionable early in the present century to identify Posidonius as the source for Latin writings in science and philosophy. H. W. Garrod's remarks (*Manilii Astronomicon Liber II* [Oxford, 1911]), p. xcix, after reading a doctoral dissertation on Posidonius as the source of Manilius are so apt and incisive that they deserve quotation: "In this laborious treatise I am not able to feel much sympathy but for the notable industry of its writer. Thinking men in Rome, necessarily, in the period in which Manilius lived, breathed in an atmosphere of Posidonius, very much as thinking men today may be said to breathe

Furthermore, discrepancies have been too much ignored.[34]

Let us consider the opinions of some scholars who have speculated on the Varronian origins of Martianus' astronomy. Pierre Courcelle decides on a Latin writer and inclines to Varro as the source of Book VIII; with good sense he does not accept the Geminus, Theon, and Cleomedes correspondences as evidence that Martianus drew upon a Greek authority.[35] Wessner pointed to the derivation, which Cassiodorus found in Varro's *De astrologia*, of *stella* from the verb *stando*.[36] Martianus (817) attributes the same derivation to a "certain Roman not unfamiliar to me"—and he surely must mean Varro. However, he goes on to attribute to Varro also the derivation of *sidera* from *considendo*; the derivation found in Varro's *De lingua latina*[37] is actually from *insidere*. Ernst Honigmann, a careful scholar and an expert in mathematical geography, examined the *climata* of Pliny and Martianus and found that whereas Pliny botches his climates, Martianus follows an Eratosthenean tradition, with an eighth climate and Roman names added; Honigmann concluded that Varro was the Latin source for Martianus' climates.[38] Pierre Duhem felt it likely that Martianus derived his astronomy from Varro, in particular his doctrines about the heliocentric motions of Venus and Mercury.[39] Sir Thomas Heath thought that Martianus' figure for the apparent diameter of the moon was probably quoted from Varro.[40] Schiaparelli, an eminent Italian astronomer and an assiduous scholar in the history of astronomy, collected forty-six passages in Greek and Latin on geoheliocentric doctrines and

---

in an atmosphere of Darwin. But not everyone who says 'Evolution' has read *The Descent of Man*; and as soon as we begin to think of Manilius as writing with a Posidonius open in front of him we make both him and ourselves ridiculous. These influences are more subtle than the critics who try to trace them."

[34] To take an example from Martianus: In his listing of constellations and tracing of celestial parallels, there is close correspondence to Aratus some of the time and absence of it at other times. This would seem to indicate a conflation of traditions. For correspondences, see below, pp. 179-80, 182-85.

[35] Courcelle, *Lettres grecques*, p. 199.

[36] Wessner, col. 2011.

[37] Varro *De lingua latina* 7. 14.

[38] *Die sieben Klimata und die* ΠΟΛΕΙΣ ΕΠΙΣΗΜΟΙ, pp. 50-52.

[39] *Le système du monde*, III, 52.

[40] *Aristarchus of Samos*, p. 314.

felt that Martianus obtained the material for Book VIII from Varro.[41] Eyssenhardt, who of all the editors of Martianus' text offered the lengthiest discussion of sources, also concluded that Varro was the source of Book VIII.[42]

The best reason for supposing that Martianus got his astronomy from Varro has been overlooked, namely, the character of the book as a whole. It is in several respects unique in Latin astronomical literature; in its orderly arrangement of topics, sense of proportion, and generally professional style of presentation, it is a gem in comparison with other Latin cosmographical treatises. Pliny is diffuse and disorganized, and often preposterous in his efforts to impress; Macrobius shows very limited knowledge and, like Pliny, makes bad blunders when he introduces digressions intended to awe the reader. Calcidius is the most professional of all the extant Latin writers of antiquity; but the astronomical portion of his commentary is largely translated from a Greek commentator on Plato,[43] and, being keyed to passages in the *Timaeus*, is not a comprehensive manual on astronomy.

We have to look to the Greek popular handbooks on astronomy for a treatise that provides an adequate basis for comparison with Book VIII of Martianus. The *Introduction to the Phaenomena* of Geminus (1st cent. B.C.) is the most conventional of the Greek treatises and serves our purposes best. Both Geminus and Martianus present systematic treatments. Though Geminus is of course far more expert, Martianus' book has the over-all characteristics of a Greek manual. Both writers define their terms, establish classifications, and expatiate on the terms defined earlier. Martianus makes free use of Greek technical terminology without transliterating it. It is clear that the ultimate forebear of Martianus' book was some popular Greek introduction to the subject. Who but Varro is likely to have been the first introduce such a handbook to Latin readers? And even if he was not the first,[44]

[41] *I precursori di Copernico nell'Antichità*, p. 27.

[42] In his 1866 edition, pp. lvi-lviii.

[43] Pseudo-Aristotle Περὶ Κόσμου.

[44] The name of Nigidius Figulus frequently crops up in this and other connections as an early authority in popular science and occult lore. For an extended treatment of Nigidius as an encyclopedist see Leonardo Ferrero, *Storia del pitagorismo nel mondo romano*, pp. 287-310.

all indications point to him as the most respected authority for Latin writers in this field.

BOOK IX: ON HARMONY

In considering Martianus' sources for his book on harmony we return to the situation observed for Books VI and VII; the very hazy figure of Varro again looms in the background but there is also a text of substantial length by another author which compares closely with Martianus' text. The greater part of Martianus' account of harmonic theory and metrics comes directly or indirectly from Book I of a Greek handbook by Aristides Quintilianus, an author who lived in the middle or later period of the Roman Empire.[45] This account is preceded by a shorter one, consisting of a general introduction and a superficial treatment of subjects which Martianus deals with more fully later. Drawn from different sources the two accounts are discrepant and repetitious.

The portion beginning at section 936 of *The Marriage* and running for thirty-four pages in the Teubner edition is a reasonably close translation of Chapters V to XIX of Book I of Aristides' *De musica*, except for many gaps where material was omitted. The text provides our best opportunity in the quadrivium books to compare Martianus' work with a Greek source. The question immediately arises of whether Martianus or some intermediate writer prepared the translation of Aristides. It is a wooden translation, with some bad misconstructions and frequent inaccuracies, which indicate that the translator did not fully comprehend the material. As Deiters points out,[46] there are paraphrases, transpositions, and interpolations of matter not found in Aristides. Yet Aristides' text can be used to fill several of the lacunae in the Martianus manuscripts, and the translation is no worse than others

[45] A new edition of Aristides' work, by R. P. Winnington-Ingram, appeared in 1963 (Leipzig: B. G. Teubner). Winnington-Ingram inclines (p. xxii) to date Aristides no earlier than the latter part of the second century, and no later than Martianus.

[46] Hermann Deiters, *Über das Verhältnis des Martianus Capella zu Aristides Quintilianus*, pp. 3-6. R. Westphal, *Die Fragmente und die Lehrsätze der griechischen Rhythmiker*, pp. 47-63, offers the parallel texts of Martianus and Aristides.

prepared by Latin compilers of the period—Calcidius and Macrobius, for example. Though he had misgivings about the relationship between Martianus and Aristides, Deiters believed that Martianus himself translated Aristides and that either author's text could serve in emending the text of the other. He attempted emendations of Aristides' text in twenty-four places. His emendations were criticized by Rudolf Schäfke, a German translator of Aristides who felt that Deiters was more successful in the reverse process, that is, in emending Martianus' text from Aristides'.[47] Professor Winnington-Ingram, the latest editor of Aristides and an outstanding expert in the field of Greek music, finds that Martianus rarely throws any light upon Aristides' text, and he is not persuaded by Deiters that Martianus had before him a copy of the text of Aristides.[48]

Of the two parts of Martianus' book on harmony the first contains a conventional introduction which enumerates instances of the supernatural and mystical powers of music. Deiters cites parallels for several of these instances from Censorinus, Gellius, Athenaeus, Pliny, Seneca, Cassiodorus, and Isidore.[49] Martianus gives Varro as his authority for one of the testimonials to the powers of music (928) and in including the designation of him as "a recent reporter" indicates that the tradition had been handed down intact through the centuries. Varro would be as good a guess as any for the main source of the early part of Martianus' book on harmony.

[47] Aristides (tr. Schäfke), p. 5.

[48] Aristides (ed. Winnington-Ingram), p. xxii.

[49] Deiters, p. 4. On p. 21 Deiters concludes that Varro was probably the source for Martianus' introductory discussion of music.

Influence

TO MODERN CRITICS Martianus has been a bundle of paradoxes. None of his failings and foibles has escaped their scorn. As an expositor of the liberal arts he shows meager skills and talents: his presentation of the trivium and quadrivium subjects is too elementary to bear favorable comparison with the more technical manuals of Donatus, Priscian, Calcidius, and Boethius. As a stylist he is outlandish. If "barbaric" is too harsh a word to describe his style, "difficult" is too mild. His absurd and bizarre allegorical setting, with its occasional lascivious touches, is a transparent device to make his book appeal to readers. To compensate for inadequacies in his ability to describe and comprehend, he resorts to grandiloquence, abstraction, and obscurity. All these strictures are valid. Yet the fact remains that Martianus was one of the half-dozen most popular and influential writers of the Middle Ages.[1]

The paradoxes, as has been suggested, exist in the minds of the

[1] Affirmations of the major importance of Martianus in the intellectual life of the Middle Ages are legion. Some were recorded above in the chapter "The Work" (esp. nn. 3, 5, and 6). Three other opinions may be cited in passing: Moses Hadas, *Ancilla to Classical Reading* (New York, 1954), p. 100, says that *The Marriage* "was among the half dozen most widely circulated books in the Middle Ages." At the opening of the chapter on "Medieval Schools to *c.* 1300" in the *CMH*, V, 765, Margaret Deanesly remarks: "The curriculum of the imperial schools, viewed by medieval scholars through the writings of Martianus Capella, consisted of the seven liberal arts: grammar, rhetoric, dialectic, geometry, arithmetic, astronomy, music." R. R. Bolgar, *The Classical Heritage and Its Beneficiaries*, p. 55, includes *The Marriage* with Cassiodorus' *Institutiones* and Isidore's *Etymologiae* among the favorite textbooks of the early Middle Ages. But the best testimony of Martianus' importance is found everywhere in Claudio Leonardi's recent book-length census of Martianus manuscripts. He points out in his introduction (pp. 467-68) that the separate disciplines of Martianus are found bound into codices containing treatises of those disciplines by other authors. This census has been of great help to me in preparing this chapter on Martianus' influence, which could easily have been expanded into a book. In collecting materials on Martianus, I have found that by far the greatest number of items belong under the heading "Influence."

critics. The traits for which they have upbraided Martianus are the very clues to his popularity. He put all seven arts in a single volume and presented them in a condensed form that was manifestly acceptable and satisfying to medieval readers. Most of them must have used the book as a review manual, to cover the ground of their secondary-school studies. His romantic setting was also an obvious success. As noted above, he became one of the key figures in chantefable in medieval literature; and his influence upon imagery in literature, painting, and sculpture was perhaps the greatest of all the Latin allegorizers.[2] Even his style, his strained conceits and neoteric diction, created a vogue.

EARLY MIDDLE AGES

During the first two centuries of its existence *The Marriage of Philology and Mercury* was used as a textbook in North Africa, Italy, Gaul, and Spain. Other recent treatises on the disciplines were also highly esteemed in the schools at this time: Boethius' manuals on each of the quadrivium subjects, Priscian's and Donatus' books on grammar, and Cassiodorus' and Isidore's concise introductions to all seven disciplines. The popularity of these competing texts may partly explain the small number of references to Martianus' book during the centuries immediately following its dissemination.

The earliest writer to cite Martianus, and perhaps the first to compose a commentary on *The Marriage*, was Fabius Planciades Fulgentius,[3] a mythographer whom Courcelle inclines to identify with St. Fulgentius (the bishop of Ruspe)[4] and whose *floruit* is usually placed

[2] See above, p. 23.

[3] *Expositio sermonum antiquorum* 1. 45 (ed. Helm, 123. 4-6): *unde et Felix Capella in libro de nuptiis Mercurii et Philologiae ait....* The title of a commentary on the first two books is listed in *Mittelalterliche Bibliothekskataloge Deutschlands und der Schweiz; herausgegeben von der Bayerischen Akademie der Wissenschaften, München*, Vol. II (Munich, 1928), p. 16.3: *Item commentum solempne Fulgencii insignis viri super duobus libris Marcialis* [sic] *de nupciis Mercurii et philologie.*

[4] Courcelle, *Lettres grecques*, p. 206. The identification has been a matter of much dispute. See Schanz, Vol. IV, pt. 2, p. 205. M. L. W. Laistner ("Fulgentius in the Carolingian Age" [1928], reprinted in *The Intellectual Heritage of the Early Middle Ages*, ed. by C. G. Starr [New York, 1966], p. 203) and M. R. P.

about the close of the fifth century. The allegorical setting of Fulgentius' *Mythologies* was inspired by Martianus, and his style shows definite traces of Martianus' influence.[5]

The next notice we have of Martianus' book is a most interesting one copied in a subscription to Book I in twenty manuscripts and to Book II in three others, according to Leonardi's census.[6] The subscription, written in the first person, records the fact that a rhetorician, Securus Melior Felix, was making a corrected copy of the text "from most corrupt manuscripts" during the consulship of Paulinus. Felix has been identified as the last person to have held the official chair of rhetoric, first occupied by Quintilian, at Rome; and the date, as 534.[7] Nevertheless, as Leonardi points out, Felix' statement is an exceedingly rare testimony regarding the editing of a text and the condition of manuscripts in antiquity.

Although Boethius is naturally associated with Martianus for having influenced medieval allegory and chantefable, there is little prospect of finding definite traces of Martianus' influence upon Boethius. James Willis, as pointed out earlier, thinks that Boethius may have glanced at the layout of Martianus' book.[8] We would not expect to find traces of Martianus' influence upon Boethius in the quadrivium books. Boe-

---

McGuire, *Introduction to Mediaeval Latin Studies*, p. 42) incline not to identify the saint and bishop and the mythographer as the same person.

[5] Willis, pp. 16-18, has collected some of the parallel passages. The Helm edition of Fulgentius has an *index sermonis*, pp. 197-215, which will serve with the Dick index to Martianus for comparisons of vocabulary.

[6] Leonardi, "I codici" (1959), pp. 444-45, and index (1960), p. 515.

[7] On the identification and date see H. I. Marrou, "Autour de la bibliothèque du Pape Agapit," *École française de Rome, Mélanges d'archéologie et d'histoire*, XLVIII (1931), 157-65. Felix is also known from his participation in a recension of Horace in 527. See Manitius, I, 7-8, n. 9; Schanz, Vol. IV, pt. 2, p. 170. The text of the subscription, as transcribed by Leonardi, "I codici" (1959), p. 446, follows: *Securus Melior Felix, vir spectabilis, comes consistorii, rhetor urbis Romae, ex mendosissimis exemplaribus emendabam, contra legente Deuterio scolastico discipulo meo, Romae, ad portam Capenam, consulatu Paulini viri clarissimi, sub die nonarum martiarum, Christo adiuvante.*

[8] Willis, p. 20, however, points to one bit of negative evidence: Boethius, in his *De institutione musica* 1. 26, says that Albinus' rendition into Latin of Greek names for notes was an innovation; but such Latin names are found in Martianus. According to Willis, Boethius either was unaware of Martianus' use of Latin

thius was translating or digesting Greek technical manuals, and his level of competence was distinctly higher than Martianus'; Martianus' direct sources were very likely all Latin. Parker rejects the suggestion that Boethius imitated Martianus' setting and prose-verse medley when he composed the *De consolatione philosophiae*; rather Parker feels that both writers imitated Varro's *Menippean Satires*.[9] Parker also feels that Boethius would have been "disgusted" by the style and matter of Martianus' book.

Cassiodorus twice refers to a book on the seven disciplines by Felix Capella but says that he has been unable to procure a copy of it.[10] Parker surmises that Cassiodorus despised Martianus' book, from what he knew of it, and tried to banish it from circulation by writing his own book of the disciplines.[11] Be that as it may, there is no question that Cassiodorus' *Institutiones* had the opposite effect. Cassiodorus' sanction of the seven secular arts in a clerical curriculum contributed greatly to the popularity of Martianus' book in Christian schools.

Reference to Martianus is also made in the work of Gregory of Tours. At the close of his *History of the Franks* Gregory addresses a fervent plea to clergymen who read his history, however schooled in the seven arts they may be, not to be impelled by its rude style to tamper with the text, to omit portions, or to destroy the book, but to transmit it whole:[12] "For even if our Martianus has instructed you in the seven arts, taught you to read in his Grammar, to recognize propositions of debate in his Dialectic, to identify meters in his Rhetoric, to calculate measurements of lands and lines in his Geometry, to trace the courses of the stars in his Astronomy, to learn the parts of numbers in his Arithmetic, and the modulations of sweet song in his Harmony—even so, I pray, do not excise anything I have written." Laistner assumes that Gregory was familiar with Martianus' book as a standard

---

terminology or deliberately concealed his knowledge, and Willis prefers the first alternative. On the antecedents and influence of Boethius' *Consolation* see Courcelle, *La Consolation de philosophie dans la tradition littéraire: Antécédents et postérité de Boèce* (Paris, 1967).

[9] Parker, p. 453.

[10] *Institutiones* 2. 2. 17; 3. 20.

[11] Parker, pp. 437, 452-53, 456.

[12] *History of the Franks* 10. 31.

school text on the liberal arts, "even if he had not studied it profoundly," and cites other references to Martianus in the *History of the Franks*.[13] Willis points out that Gregory was the first one to call Martianus by that name, and also that Dick, in listing medieval *testimonia* in his 1925 edition, confused Gregory of Tours with Gregory the Great.[14]

Evidence of Martianus' influence has been observed in an anonymous composition of Welsh or Irish provenance entitled *Hisperica famina*. This work, dating from the second half of the sixth century, numbers more than six hundred nonmetrical lines, which display marked assonances. Most of the lines consist of two parts, the first part containing one or two epithets, the second part containing a subject and verb. The work is noteworthy for its strange Latinity.[15] Willis finds in the vocabulary "strong affinities" with Martianus,[16] and Manitius points to other signs of borrowing from Martianus.[17]

The borrowing from Martianus by Isidore of Seville in his encyclopedic *Etymologies* was "cautious but sustained" (*discret mais suivi*), according to Jacques Fontaine,[18] who believes that Isidore made direct use of *The Marriage*.[19] Leonardi does not consider that the parallels cited by Fontaine are conclusive evidence of direct use. He prefers to

[13] *Thought and Letters in Western Europe A.D. 500 to 900*, p. 129. Gregory certainly did not study Martianus profoundly, as his quoted statement shows. One would not learn to read by studying Martianus' *De grammatica*, and the identification of meters is found in the *De harmonia*, not the *De rhetorica*. Erich Auerbach, *Literary Language and Its Public in Late Latin Antiquity and in the Middle Ages*, pp. 106-9, discusses the tone and content of this passage from Gregory and interprets it in the light of what is known about his style. See also Leonardi, "I codici" (1959), pp. 459-60; T. J. Haarhoff, *Schools of Gaul* (Johannesburg, 1958), pp. 188-89.

[14] Willis, p. 21. Cf. Dick ed., p. xxix.

[15] The term "Hisperic Latin" is applied to its style of composition, as well as to the style of other similar works: the *Lorica*, attributed to Gildas; and the *Altus Prosator*, attributed to St. Columba. See *The Cambridge History of English Literature*, ed. A. W. Ward and A. R. Waller (Cambridge, 1960), I, 69.

[16] Willis, p. 24.

[17] Manitius, I, 156-58. See also F. J. H. Jenkinson, *The Hisperica Famina* (Cambridge, 1908) for the text and an *Index verborum*.

[18] *Isidore de Séville et la culture classique dans l'Espagne wisigothique*, II, 858; and see *Index Locorum*, pp. 969-70.

[19] See *ibid.*, esp. I, 352, n. 1; 442; II, 748, 749, n. 1.

believe that Isidore possessed only Book VII of Martianus' work and that he used this in compiling the *Liber de numeris.*[20] Following Barwick's suggestion—in his review of Dick's edition of Martianus—that the text of Book VII might be corrected in several places by comparisons with Isidore's text,[21] Leonardi sets up the parallel passages and offers emended readings. He also shows how Isidore's text may be emended by comparisons with Martianus[22] and makes a strong case for his contention that the texts of compilations derived from Martianus are of primary importance in improving his text. Fontaine demonstrates parallels in phraseology between Martianus and a poem which was dedicated to Isidore by King Sisebut and appended to the text of Isidore's *De natura rerum.*[23]

Considering that Martianus was to be a favorite author of the ninth-century Irish scholars who emigrated to Charlemagne's Empire, it is remarkable that so few traces of his work have been found from England during the early Middle Ages. And while we are reflecting upon long-range influences, it may not be amiss to point out that Martianus has never been a popular author in the British Isles.[24] We

[20] Leonardi, "I codici" (1959), p. 461, n. 94; and "Intorno al 'Liber de numeris,'" p. 231.

[21] *Gnomon*, II (1926), 190.

[22] "Intorno al 'Liber de numeris,'" pp. 224-27 for emendations of Martianus' text, pp. 227-31 for emendations of Isidore's text. Fontaine also seems to feel greater assurance in dealing with Martianus' borrowings in Isidore's *Liber de numeris*; see Fontaine, I, 380; 371, n. 4; 375; 399; 401, n. 1; 402. On Isidore's use of Book VII see also Manitius, I, 57-58. It should be pointed out that some scholars have recently denied Isidore's authorship of the *Liber de numeris.* See J. N. Hillgarth, "The Position of Isidorian Studies," in *Isidorians*, pp. 23-24.

[23] *Isidore de Séville Traité de la nature*, ed. J. Fontaine, pp. 156, 328-29, 332-33, and 334-35.

[24] Rarely do we find the neologisms of Martianus, listed in Appendix B, cited in R. E. Latham, ed., *Revised Medieval Latin Word-List from British and Irish Sources* (which covers works into modern times). J. D. Ogilvy, *Books Known to the English, 597-1066*, p. 4, observes that "Macrobius' *Saturnalia* and Martianus' *De nuptiis*, though known in Ireland in the seventh century, were unknown to Bede; and there is no evidence that they reached England before the tenth century." C. W. Jones, in his *Bedae Opera de temporibus*, p. 10, flatly states that Martianus was not known to Bede. R. C. Jebb ("The Classical Renaissance" in *The Cambridge Modern History*, I, 533) points out that Martianus' book was not listed in Alcuin's catalogue (*c.* 770) of the library at York and attributes the ab-

may even see a connection between present-day British eschewal of the writings of classical pedants like Martianus—for the past century such writings have provided Continental classical philologists with a generous store of dissertation subjects—and the neglect of Martianus in England during the Middle Ages.

Willis finds "slight evidence" of Bede's use of Martianus,[25] but the instances he cites are admittedly inconclusive and in the example of parallelism that he quotes from Manitius, Bede is actually echoing Pliny rather than Martianus.[26] The authors known to have been Bede's sources for his treatises on cosmography and chronology were Pliny, Isidore, and Macrobius. As for other possible influences in this period, Raby thinks that Martianus was the model for a poem in Adonic verse addressed by the Irish scholar Columban to his friend Fidolius.[27]

CAROLINGIAN AGE

In the third generation of the Carolingian dynasty Martianus' book came into a popularity such as it had not experienced during the early Middle Ages,[28] and it continued to hold a prominent place in intellectual and literary developments throughout the later Middle Ages. To trace and analyze Martianus' manifold influences from this time on

---

sence of a copy there to the hostility of Christian teachers toward pagan literature. York had at the time the best collection of books in northwestern Europe.

[25] Willis, p. 25.

[26] Manitius, I, 77, says: "anklingend Mart. Cap." But cf. Bede *De natura rerum* 3 (Migne, *PL*, Vol. XC, col. 192): *Mundus est universitas omnis quae constat ex caelo et terra, quattuor elementis in speciem orbis absoluti globata*; Pliny *Natural History* 2. 5, 8, 10: *Formam eius in speciem orbis absoluti globatam esse . . .*; *nos* [*appelavimus*] *mundum; nec de elementis video dubitari quattuor esse ea*; Martianus 814 (Dick 430. 12-13): *Mundus igitur ex quattuor elementis isdemque totis in sphaerae modum globatur.*

[27] *Secular Latin Poetry*, I, 164.

[28] Leonardi, "I codici" (1959), pp. 448-49, sees the "rediscovery" (*riscoperta*) of Martianus as intimately related to Carolingian intellectual life. He remarks (p. 462) that of the more than fifty extant codices containing the unabridged text of the *De nuptiis* nearly half are to be dated in the ninth or tenth century, and nearly all of these manuscripts have abundant glosses. See Laistner, *Thought and Letters*, pp. 213-15, on Martianus' place in Carolingian education.

would require a book-length study.[29] Cappuyns remarked upon the surprising omission of Martianus' name in the literature of the early Carolingian Age,[30] but Willis points to a commentary[31] on Martianus earlier than that of John Scot Eriugena and to reminiscences in the poems of Walafrid Strabo as evidence that Martianus was studied and read during that period.[32]

A poem that was to become immensely popular and was to be used as a textbook in medieval schools from the eleventh century on is the *Ecloga Theoduli*, of uncertain authorship. Manitius, favoring Osternacher's suggestion, inclines to attribute the poem to Godescalc (Gottschalk);[33] Raby agrees that the poem belongs to the ninth century but thinks that the author is still unknown.[34] The mythological material in the poem was drawn from Vergil, Ovid, Servius' *Commentary*, and Martianus Capella.[35]

The unknown author of a geographical treatise entitled *De situ orbis* and compiled shortly after 850 transcribed the greater part of the text of the sixth book of Martianus' work (sections 617-703, with gaps), though he used other sources as well.[36] Dick made no use of the *De situ orbis* (edited by Manitius) in his 1925 edition, evidently because he was unaware of its existence, and was later taken to task by Manitius[37] for not having included it in his collation. As Manitius observed,

[29] The exhaustive studies of Max Manitius in tracing influences of classical authors in the Middle Ages make it unnecessary to offer here anything more than a selective survey. See the indexes to the three volumes of his *Geschichte der lateinischen Literatur des Mittelalters*. For authors, titles, and some bibliography on Martianus' *Fortleben* see also Wessner, cols. 2012-13; Cappuyns, "Capella," cols. 844-46; Schanz, Vol. IV, pt. 2, p. 170.

[30] M. Cappuyns, *Jean Scot Érigène*, p. 79.

[31] Formerly referred to as the "Dunchad" commentary, now attributed to Martin of Laon. See n. 42 below.

[32] Willis, pp. 26-28.

[33] Manitius, I, 570, 572-73.

[34] *Secular Latin Poetry*, I, 228.

[35] Manitius, I, 573.

[36] Leonardi, "Nota introduttiva," p. 276, lists the parallel passages. On the sources of the *De situ orbis* see Manitius, I, 675-76.

[37] In his review of the Dick edition in *Philologische Wochenschrift*, XLV (1925), 543. Manitius' edition of the *De situ orbis* had appeared in 1884.

the Martianus manuscript used by the anomymous compiler was older than any now in existence.

The Marriage of Philology and Mercury became one of the leading textbooks in the schools of this period, owing largely to the interest drawn to it by the renowned Irish scholars who emigrated to the Frankish lands in the ninth century—John Scot Eriugena, Martin of Laon, and Remigius of Auxerre.[38] John Scot, who depended heavily on Martianus in cosmology,[39] lectured on him at the Palace School; Martin taught at the Irish colony at Laon; and Remigius taught at Auxerre. Their commentaries on Martianus, comprising large groups of glosses not yet clearly identified, were published for the first time by Cora E. Lutz.[40] Since then, opinions have changed about the nature of John's glosses.[41] And Jean Préaux presents cogent arguments for

[38] Remigius' birthplace and his whereabouts before he came to Auxerre have been matters of speculation. Cora Lutz (Remigius, I, 5) in saying that he was born "probably of Burgundian parents" relies upon Manitius' observation (I, 504) that Remigius evidently understood the Frankish language. (The evidence for this is the meaning of a single word, *Hungri: a fame quem patiebantur, Hungri vocati sunt.*) Greater reliance may be placed on an autobiographical remark, heretofore overlooked, which is made by Remigius in his *Commentary* (ed. Lutz, II, 269): "in our climate a solstitial day has eighteen hours of daylight" (*secundum nostrum clima X et VIII horas habet dies solsticialis*). This appears to be a datum about Ireland that he learned in school. Classical geographers recorded seventeen hours of daylight for northern Britain. Laistner, *Thought and Letters*, p. 214, seems to call Remigius an Irishman, but he does not give his evidence.

[39] See Liebeschütz, pp. 127-37. The studies of E. von Erhardt-Siebold and R. von Erhardt, *The Astronomy of Johannes Scotus Erigena* and *Cosmology in the Annotationes in Marcianum*, are useful for suggestions and bibliography but must be read with caution.

[40] *Iohannis Scotti Annotationes in Marcianum* (1939); Dunchad *Glossae in Martianum* (1944); *Remigii Autissiodorensis Commentum in Martianum Capellam* (1962-1965).

[41] Miss Lutz assumed that the manuscript that she used in her edition of John's commentary was unique. However, Lotte Labowsky four years later discovered another. E. K. Rand concluded that only Books I and II of the commentary in the 1939 Lutz edition were by John, Book II being in a condensed form. Miss Lutz now feels that Books VI-IX are abridged from the better of the extant versions of Remigius' commentary, and that John's glosses for these books, if they were prepared, are no longer extant. She also feels that the glosses for Books IV-V in their present form are reduced from John's original commentary. For a report on these and other studies of the Carolingian commentaries on Martianus

ascribing to Martin of Laon the glosses which, at the time of Miss Lutz' editing, were commonly attributed to Dunchad. Miss Lutz edited the "Dunchad" glosses for the last third of Book II, all of Book IV, and the first third of Book V. Préaux has since discovered other, more extensive glosses, for all nine books.[42] When an edition of these glosses is prepared, it will rank in importance with the commentary of Remigius. Still another anonymous commentary, found in two Cambridge manuscripts,[43] awaits publication. In the meantime the recently published commentary of Remigius serves as the fullest and best medieval explication of Martianus.[44] The keenest interest in Martianus studies in the Carolingian Age appears to have been sown in the centers of northeastern France, at Corbie, Laon, Rheims, and Auxerre.[45]

Didactic matters figure prominently in several published poems[46] from the ninth to eleventh centuries which were based upon or influenced by Martianus. The trivium and quadrivium subjects are personified, and individually treated. Some of these poems were probably composed as school exercises.

---

see Hans Liebeschütz, "Zur Geschichte der Erklärung des Martianus Capella bei Eriugena," *Philologus*, CIV (1960), 127-37; and my article, "To a Better Understanding of Martianus Capella," *Speculum*, XL (1965), 107-10, 113-14.

[42] "Le Commentaire de Martin de Laon," *Latomus*, XII (1953), 437-59.

[43] Corpus Christi College Library, Nos. 153, 330. Professor Préaux has expressed in a letter to me his intention to edit the Martin and anonymous Cambridge commentaries for Books I and II only.

[44] An anonymous commentary on Books I and II which dates from the end of the twelfth century (Cod. Vat. Barb. lat. 10), has been edited by Ann Rose Raia (Unpublished Ph.D. dissertation, Fordham University, 1965). Miss Raia finds that this commentary follows Remigius closely, especially in Book I, but that it adds "a significant amount of material' (p. 340).

[45] See Leonardi, "I codici" (1959), p. 465; B. L. Ullman, p. 275. Professor Ullman argues strenuously for Corbie as the center of study of all the liberal arts during the Carolingian Age.

[46] *Monumenta Germaniae Historica, Poetarum Latinorum Medii Aevi*, Vol. I (Berlin, 1881), pp. 408-10, 544-47, 629-30; Vol. IV, fasc. 1 (1899), pp. 249-60, 339-43. Leonardi discusses these poems at length in "Nuove voci poetiche tra secolo IX e XI," *SM*, ser. 3, II, (1961), 139-68. These poems are also discussed by Laistner, *Thought and Letters*, p. 214; Curtius, p. 39; Raby, *Secular Latin Poetry*, I, 347-48; *A History of Christian Latin Poetry from the Beginnings to the Close of the Middle Ages*, 2d ed., p. 282.

POST-CAROLINGIAN PERIOD

Another revival of interest in Martianus took place in Italy in the tenth century. Leonardi points to Rather of Verona, Stephen and Gunzo of Novara, Eugenius Vulgarius, Liutprand of Cremona, and the glosses of the *Gesta Berengarii* as instances of that revival.[47] Rather is the most significant figure among these. An outstanding scholar in his day, he played an important part in the copying and editing of a group of Martianus manuscripts, and his glosses have come under close study.[48]

We would expect Gerbert (Pope Sylvester II from 999 until his death in 1003), who was avidly interested in Arabic science, to have been familiar with Martianus' quadrivium books. In Letter 161, addressed to Brother Adam and dated March 10, 989, Gerbert quotes Martianus on the amounts of increase and decrease in daylight during the various months of the year.[49]

Notker Labeo (d. 1022), one of the leading scholars of St. Gall and a most assiduous translator of classical works into Old High German, prepared a translation, with interpolated commentary, of Books I and II of *The Marriage*, the commentary being derived largely from Remigius' *Commentary*. Notker's translations have been thoroughly studied, because of their intrinsic interest as specimens of Old High German and because of Notker's outstanding position in the history of that language.[50]

[47] See Leonardi, "I codici" (1959), pp. 469-70, for the references.

[48] See *idem*, "Raterio e Marziano Capella," pp. 73-102, for a description of this group of manuscripts and the Rather glosses, and for an account of the studies of Bischoff, Weigle, Préaux, and others. Auerbach, pp. 133-52, gives a fine account of Rather's career and writings; he regards Rather as "the most important and interesting Latin writer of his time."

[49] See *The Letters of Gerbert, with His Papal Privileges as Sylvester II*, tr. Harriet Pratt Lattin, pp. 189-91. See also Gerbert's *Opera mathematica*, ed. N. M. Bubnov, pp. 39-41; Manitius, II, 668, 670; Duhem, III, 63. The passage quoted by Gerbert is one of the two most frequently excerpted passages from Martianus' Book VIII in medieval codices on astronomy. See Leonardi, "I codici" (1959), p. 482.

[50] The standard edition is *Notkers des Deutschen Werke*, Vol. II: *Marcianus Capella, De nuptiis Philologiae et Mercurii*, ed. E. H. Sehrt and T. Starck. See also A. K. Dolch, *Notker-Studien I-II*; Karl Schulte, *Das Verhältnis von Notkers De nupti is Philologiae et Mercurii zum Kommentar des Remigius Autissiodorensis*,

Martianus' emergence as a model of genre-writing began about the time of this quickening of interest in *The Marriage*. Though the Christian poet Prudentius continued to exercise the greatest influence upon Christian allegorists, Martianus, coming at the close of the Empire, enjoyed an advantage in secular allegory over earlier masters, such as Statius and Lucan,[51] because medieval readers generally preferred later exponents of learning and genre-writing to earlier ones.[52] In the graphic and plastic arts Martianus' allegorical figures appeared first in ninth- and tenth-century manuscript miniatures and gained prominence in the façades of the cathedrals at Chartres, Laon, Auxerre, and Paris.[53] In the later Middle Ages didactic poets, following the lead of Prudentius and Martianus, took to personifying most of the aspects of human thought and feeling.[54]

Again, with *prosimetrum*, it was not the early masters—Varro, who in his voluminous collection of *Menippean Satires* introduced the genre into Latin literature, or the pungent and ribald satirists Petronius and Seneca—who served as models for the form. Whatever influence Varro had upon the genre in the Middle Ages was exercised through Martianus' book and the *Consolation* of Boethius.[55] Instances of Martianus' influence are too numerous to cite here. They may be traced through the indexes to the volumes of Manitius' *Geschichte*, the bibliographical notes in Leonardi's census, and the studies of Cappuyns, Wessner, Schanz, and Raby.[56]

---

a dissertation almost entirely (pp. 3-89) devoted to comparison of the texts of Remigius and Notker in parallel columns; J. H. Tisch, "Martianus Capella and Notker Teutonicus: The Creative Challenge," *Sydney University Medieval Group Newsletter*, V (1965), 29-49; H. O. Taylor, *The Mediaeval Mind*, 4th ed., I, 309-10.

[51] These allegorists are discussed with fine perception and insight by C. S. Lewis, *The Allegory of Love*, chap. II.

[52] Curtius, p. 104, discusses at length the medieval *topos* of the young-old supernatural woman, the sort personifying the liberal arts in Martianus, and points out that the figure had already degenerated into a rhetorical cliché when Boethius gave it new life in a religious context.

[53] See Appendix A.

[54] R. E. F. Klibansky, E. Panofsky, and F. Saxl, *Saturn and Melancholy*, p. 221.

[55] On the influence of Boethius' *Consolation*, see n. 8 above.

[56] Leonardi, "I codici" (1959), pp. 469-70, 474-83; Cappuyns, "Capella," cols.

One medieval writer may be singled out for mention here as particularly significant. Alan of Lille (Alanus de Insulis), the "Universal Doctor," in drawing inspiration for the philosophical allegory and *prosimetrum* form of his *Anticlaudianus*, looked backward to Martianus and, through Dante, Chaucer, and others, helped to transmit Martianus' influence to the later Middle Ages.[57] The *Anticlaudianus* has appeared in a new edition,[58] and its sources and influences have been thoroughly investigated.[59]

SCHOLASTICS OF THE TWELFTH CENTURY

The trend among the Scholastics to replace rhetoric with science reached its full development at Chartres in the twelfth century.[60] The absorbing interest in Platonism and cosmography that characterized the school of Chartres swept the Neoplatonists and Calcidius and Macrobius to the pinnacle of their influence in the Middle Ages and gave Martianus a respect and authority in scientific matters such as he had not enjoyed since the Carolingian Age.[61]

Martianus was one of the main sources of the *De mundi universitate* of Bernard Silvestris.[62] The work is fraught with allegory and is composed in the mixed prose and verse form. Jeauneau has recently discovered a fragment of a commentary on Martianus which he attributes to Bernard.[63] The fragment, covering sections 1-37 of Book I, is found

---

844-47; Wessner, cols. 2012-13; Schanz, Vol. IV, pt. 2, p. 170; and Raby, *Secular Latin Poetry*, chap. I, sec 2; chap. X, sec. 2.

[57] An extended discussion and summary of the *Anticlaudianus* is to be found in Taylor, II, 120-30. See also C. S. Lewis, *The Allegory of Love*, pp. 98-105.

[58] Critical text and translation with tables by R. Bossuat (Paris, 1955).

[59] See Curtius, pp. 117-21, 360-61; Raby, *Christian Latin Poetry*, p. 300; F. Schalk, "Zur Entwicklung der Artes in Frankreich und Italien," in *Artes Liberales von der antiken Bildung zur Wissenschaft des Mittelalters*, ed. Josef Koch, pp. 143-44.

[60] Auerbach, pp. 274-77; Haskins, *Studies in the History of Mediaeval Science*, pp. 88-92.

[61] See Jeauneau, pp. 842-43.

[62] See Theodore Silverstein, "The Fabulous Cosmogony of Bernardus Silvestris," *Modern Philology*, XLVI (1948), 96-116; Raby, *Secular Latin Poetry*, II, 8-13; Curtius, pp. 108-13.

[63] Jeauneau, pp. 844-51.

on folios 1-28 of a Cambridge manuscript[64] dated in the first half of the thirteenth century. Jeauneau has transcribed portions of the commentary.[65]

Another allegory inspired by Martianus and the *Consolation* of Boethius is the *De eodem et diverso* of Adelard of Bath. Philosophy, representing permanence, wins a debate with Love-of-the-World (*Philocosmia*), representing change and decay, and goes on to explain the purpose of the seven liberal arts, personifications of which surround her.[66]

Also influenced by Martianus was the *Eptateuchon* of Thierry of Chartres, a large encyclopedia of the seven liberal arts, comprising almost six hundred folios. Thierry admits in his preface that the work is not his own but is compiled from "the chief authorities on the arts." The authorities he names are Varro, Pliny, and Martianus. His book represent a wedding of the trivium (Philosophy) to the quadrivium (Mercury), and he attributes this familiar setting of Martianus to "the Greek and Latin poets."[67] The work has never been published. According to Clerval, it transcribes Books V and VII of Martianus in their entirety.[68]

Another unpublished commentary on Martianus has been doubtfully attributed to William of Conches.[69] Still another commentary on

[64] University Library, Mm. 1. 18.

[65] Jeauneau, pp. 855-64.

[66] See *Des Adelard von Bath Traktat de eodem et diverso*, ed. with introduction and commentary by Hans Willner (Münster, 1903), p. 38; Haskins, *Studies in the History of Mediaeval Science*, chap. II.

[67] For a translation of the preface of the *Eptateuchon* see Paul Vignaux, *Philosophy in the Middle Ages* (London, 1959), p. 29.

[68] J. A. Clerval, "L'Enseignement des arts libéraux à Chartres et à Paris dans la première moitié du XIIe siècle, d'après l' *Heptateuchon* de Thierry de Chartres," in *Congrès scientifique international des Catholiques tenu à Paris en 1888*, II (Paris, 1888), 281-82. For recent studies of the *Eptateuchon* see Jeauneau, pp. 829-30, 853-54; Leonardi, "I codici' (1959), p. 476. The unique copy of the *Eptateuchon*, formerly found in MSS 497 and 498 of the municipal library at Chartres, was destroyed in a bombardment of May 26, 1944. Fortunately microfilm copies had been made—one for Mont-César, Louvain; another for the Pontifical Institute of Mediaeval Studies, Toronto—and other copies have since been distributed in Europe, America, and Australia. See Jeauneau, p. 853.

[69] By M. Grabmann, in "Handschriftliche Forschungen und Mitteilungen zum

Martianus (Books I and II only) by Alexander Neckam is being edited for the first time by R. A. van Kluyve of Duke University and Judson Allen of Wake Forest College.[70] Of one hundred and fifty miscellaneous Barberini codices (Vatican Library) recently catalogued and described by Sesto Prete, three (Barb. lat. 5, 10, and 130) are to be found in Leonardi's census. A fourth, overlooked by Leonardi, is a short excerpt from Martianus.[71]

That John of Salisbury was familiar with *The Marriage of Philology and Mercury* is seen in his numerous references to the work in his *Metalogicon*.[72] Martianus was one of the main sources used by the geographers Adam of Bremen[73] and Lambert of St. Omer,[74] and of the *De proprietatibus rerum* of Bartholomew of England.[75] In the poem the *Fons philosophiae*, by Godfrey of St. Victor, the poet takes an imaginary journey on which he turns away from the foul waters of the seven mechanical arts and meets Martianus among the masters of wisdom.[76]

Martianus, Macrobius, and Calcidius, the revered authorities on cosmography at Chartres, were the three late Latin authors who were largely responsible for keeping alive in the Middle Ages the belief in

---

Schrifttum des Wilhelm von Conches," *Sitzungsberichte der Bayerischen Akademie der Wissenschaften, Philosophisch-historische Abteilung*, 1935, pt. 10, pp. 25-26. Regarding this commentary see Jeauneau, pp. 832-34; Leonardi, "I codici" (1959), p. 476.

[70] The editors informed me, in a letter dated March 17, 1966, that they have transcribed Bodleian Digby 221 and are collating it against Cambridge Gonville and Caius 372/621. On Neckam's commentary see Jeauneau, p. 830.

[71] Barb. lat. 92, fols. 45v-46r. See *Bibliothecae Apostolicae Vaticanae codices manu scripti recensiti: Codices Barberiniani Latini 1-150 recensuit Sextus Prete* (Rome, 1968), pp. 8, 13-14, 165, 237-38. Barb. lat. 10 was edited, under the direction of Professor Prete, by Ann Rose Raia. See n. 44 above.

[72] See *The Metalogicon of John of Salisbury*, tr. Daniel D. McGarry (Berkeley, 1962), Index.

[73] Beazley, II, 522, 530, 542, 546.

[74] *Ibid.*, pp. 571-72; 622-23.

[75] See H. C. Darby, "Geography in a Medieval Text-Book," *Scottish Geographical Magazine*, XLIX (1933), 324; F. R. Johnson, *Astronomical Thought in Renaissance England: A Study of the English Scientific Writings from 1500-1645* (Baltimore, 1937), p. 72.

[76] Raby, *Secular Latin Poetry*, II, 13.

a spherical earth and in the geoheliocentric orbits of Venus and Mercury. Historians of astronomy regard the impact that these writers had upon medieval philosophers as a significant factor in the evolution of the Copernican theory. Duhem devotes a chapter of 119 pages[77] to tracing the transmision of the geoheliocentric theory ("Heraclidean" system) in the Middle Ages.[78] Martianus and Macrobius were also responsible for transmitting to medieval geographers Eratosthenes' estimate of 252,000 stadia for the earth's circumference, and Crates' (2d cent. B.C.) conception of four inhabited quarters of the globe, diametrically or transversely opposed to each other in the northern and southern or eastern and western hemispheres.[79]

LATER MIDDLE AGES

During the later Middle Ages, interest continued to be strong in the setting portion (Books I-II) of *The Marriage*, but the discipline books generally declined in popularity. The one exception was Book VIII, on astronomy, which was often excerpted, either from the beginning of the discipline proper (814: *Mundus igitur ex quattuor elementis* . . .) to the end of the book,[80] or in smaller segments, particularly sections 844-45 and 855-77.[81] Codices which comprise miscellaneous tracts and excerpts on astronomy turn up in abundance, sometimes disguised with

[77] *Le Système du monde*, III, 144-62. See also Dreyer, *History of Astronomy*, p. 207; and his "Medieval Astronomy," in *Toward Modern Science*, ed. R. M. Palter, I, 237, for the importance of Martianus in medieval astronomy.

[78] See Beazley, I, 340-43; Wright, pp. 54, 158-160, 366; G. H. T. Kimble, *Geography in the Middle Ages* (London, 1938), pp. 8-9, 11, 24. M. Cary and E. H. Warmington, *The Ancient Explorers* (Harmondsworth, 1963), p. 231, view the Cratean conception of antipodes, perioeci, and antoeci as a factor in explorations leading to the discovery of America.

[79] Leonardi, "I codici" (1959), pp. 473-74, points out that of 55 codices that contain just Books I and II, only one or two date from the ninth and tenth centuries, 32 date from the eleventh to fourteenth centuries, and 11 are from the fifteenth century—clear indication that Books I and II held their chief attraction for readers in the late Middle Ages.

[80] *Ibid.*, p. 472, lists 22 of these acephalous fragments dating from the twelfth to fourteenth centuries.

[81] *Ibid.*, p. 482.

the title *Liber Yparci* or *Iparci* ("Book of Hipparchus").[82] These frequently contain excerpts from Martianus' eighth book.

The respect and admiration that Martianus lost as a man of learning was compensated for by the growing appreciation of his allegorical figures and fanciful imagination. The heavenly journey of Philology served as a model and inspiration for other similar literary journeys, including that of Dante through the celestial spheres.[83] Petrarch owned a copy of Martianus but was influenced by him little, if at all.[84] Surprising in both its late appearance and its popularity is the encyclopedia *Visión Delectable de la Philosophía y Artes Liberales, Metaphísica y Philosophía Moral*, compiled by Alfonso de la Torre about 1435.[85] It is based upon Martianus, in its structure as well as its allegorical approach.

[82] *Ibid.*, pp. 471, 472, 486; and the indexes to Leonardi, "I codici" (1959 and 1960). See also Haskins, *Studies in the History of Medieval Science*, p. 89. Another excerpt that turns up in large numbers in the codices is the section on cosmography in Macrobius' *Commentary* (1. 14 - 2. 9). Von Jan used four of these in the collation for his 1848 edition of Macrobius. Marginal notations and rubrics are so common at the beginning and end of this excerpt in the manuscripts that von Jan noted (p. lxxv) three manuscripts that do not have these notations.

[83] On Martianus' importance as an inspiration to Dante see Curtius, p. 360. H. R. Patch, *The Other World According to Descriptions in Medieval Literature* (Cambridge, Mass., 1950), traces the heavenly journeys in medieval literature. There is need for a similar study of the classical examples, such as are found in Plato's *Republic*, the proem of Parmenides' fragmentary poem, Eratosthenes' *Hermes*, Cicero's *Dream of Scipio*, Macrobius' *Commentary*, and Martianus. R. M. Jones, "Posidonius and the Flight of the Mind Through the Universe," *CP*, XXI (1926), 97-113, discusses the important classical conceptions and is critical of those who represent Posidonius as the popular source and inspiration of later journeys.

[84] Leonardi, "I codici" (1959), pp. 455, 480.

[85] See J. P. W. Crawford, "The Seven Liberal Arts in the *Visión Delectable* of Alfonso de la Torre," *Romanic Review*, IV (1913), 58-75. See also Robert Collier, *Encyclopaedias: Their History Throughout the Ages* (New York, 1964), pp. 72-74. Achille Pellizzari surveys quadrivium studies in Italy during the Renaissance in *Il quadrivio nel rinascimento* (Naples, 1924).

Manuscripts and Editions

THE OBSCURITY of Martianus' style and diction generated corruptions in his text from the beginning. Securus Melior Felix, as we have previously observed, testified in 534 to the wretched condition of the texts he was using for the corrected copy he was making.[1] Agreement in the lacunae of the extant manuscripts of Martianus indicates that they all stem from one copy; but it would be unwarranted to conclude that Felix' copy was that archetype, since his subscription is found only at the end of either Book I or Book II.[2] Because of its serviceability as a textbook and its attractions as a compendium of classical learning, *The Marriage* was profusely copied and was excerpted by compilers and commented upon by glossators and schoolmasters for a thousand years. The earlier compilers were copying from manuscripts older than any now extant. The glaring defect of modern editions of Martianus, including the most recent, by Dick, is that insufficient use has been made of the excerpts in later compilers and the lemmata of commentators. The vast number of excerpts and lemmata that should be considered in a new collation, as well as the bad condition of the extant Martianus manuscripts, combine to make him one of the most difficult of Latin authors to edit.

When the article on Martianus was written for *The Oxford Classical Dictionary* (1949) few of the numerous manuscripts of *The Marriage* had been examined. But since then, Claudio Leonardi has examined the greater number of them, including most of the important ones, and in 1959-1960 he published a census of all the manuscripts he could trace in the libraries of Europe. His list contains 243 items, two of them being added while his census was going through press. Of these, 176 are fully described from his own examination; descriptions of 53 others depend upon library catalogues; the rest have been examined at

[1] See above, p. 57.

[2] Leonardi, "I codici" (1959), p. 446. Cappuyns, "Capella," col. 843, believes, however, that Felix' copy was the archetype.

least in part on microfilm. A melancholy note is injected into the census in that 6 of the manuscripts were destroyed in war bombardments. Leonardi's census contains the following data for each manuscript: general description, provenance, ownership, contents of the entire codex containing the Martianus text, pagination of each book of Martianus' work in the longer manuscripts, corrections and emendations introduced into the text, bibliography of existing catalogue listings and descriptions, and references to scholarly studies of the manuscripts.

Two important observations result from Leonardi's census and his other studies on the *fortuna* of Martianus: First, from the earliest manuscripts on, the very number of the complete copies and of the books and passages that were excerpted in each century indicate the course of intellectual history in the Latin West. To illustrate: The abundance of complete manuscripts from the ninth and tenth centuries and the large number of glosses that accompany those manuscripts testify to the heightened interest in the seven liberal arts during the Carolingian revival. The largest number of manuscripts of Book VIII as a separate excerpt, with the greatest profusion of glosses and diagrams, appears in the twelfth century, at the time when the Scholastics of Chartres were keenly interested in cosmography. The abundance of manuscripts dating from the twelfth to fourteenth centuries which contain only Books I and II supports the assumption that Martianus' learning was no longer able to withstand the competition of better manuals of the disciplines emerging in translated form from the Greco-Arabic transmission, and also indicates that Martianus had found a new popularity as an allegorist and romancer. Second, because of the widespread copying from Martianus' book by later compilers who were using better manuscripts than those we possess, the likeliest prospects for improving Martianus' text now lie in collating the readings of later excerpts. Not all editors agree with Leonardi's second contention, however.[3]

[3] I incline to favor Leonardi's view. Remigius' bulky commentary on Martianus has just been published, and its use of Martianus' words and phrases in lemmata, together with those excerpts found in commentaries published earlier, must be taken into account by textual critics. It is hoped that still other commentaries will be published. To cite a single instance of the importance of the readings in the

Dick's edition was, according to the consensus of reviewers and critics, far superior to its predecessors. Its two main defects, as reported by those reviewers and critics, were its restricted collation, omitting several of the best manuscripts, and its very limited use of passages and phrases quoted by later compilers. These defects should be avoided by future editors and textual critics. Dick used thirteen manuscripts in his collation, limiting himself to certain libraries in Germany, Switzerland, and the Netherlands, and overlooking important manuscripts in libraries in England, Paris, the Vatican, and elsewhere. Recent studies by Jean Préaux, Bernhard Bischoff, Claudio Leonardi, and others have clearly indicated the importance of the manuscripts overlooked by Dick and the need for a new edition.[4]

It is surprising, when one considers Martianus' popularity in the Middle Ages, that his work appeared in relatively few printed editions and that the princeps edition (1499) appeared so late. Macrobius' works, for example, experienced precisely the same *fortuna* as Mar-

---

commentaries for emending Martianus' text, I quote from Percy Jones, *The Glosses* De Musica *of John Scottus Eriugena in the MS. Lat. 12960 of the Bibliothèque Nationale, Paris*, p. 88: "A comparative study of the differences of readings between Eriugena's text and of Dick's edition indicates that there are no copies of the manuscripts used by Eriugena now in existence. In fact there are sixty-three variants of Martianus' texts as given by our commentator which are not used by Dick in his edition. In any future edition of Martianus, Eriugena's commentary should be used as an essential aid in the reconstruction of the original text. It is more than a hundred years older than the earliest manuscript of the *De nuptiis* which we have today and a study of its readings as against those of Dick shows, in the vast majority of these variants, even those manuscripts which are still in existence, are in agreement with Eriugena's text and not with that of Dick. The obvious conclusion from the figures [omitted here] is that we have in Eriugena's commentary a far more correct reading than in Dick's restored version." Moreover, Barwick, Manitius, and Wessner criticized Dick for not availing himself of readings in the texts of later compilers. Leonardi undertook the emendation of the text of Martianus' Book VII from the *Liber de numeris* of Isidore and from the Rather glosses. One experienced editor, however, has indicated to me in a letter that he does not feel that the *fortuna* is more important than the Martianus manuscripts in emending present texts.

[4] Préaux' studies are reported on in my article, "To a Better Understanding of Martianus Capella," pp. 113-14. References to the studies of Bischoff, Leonardi, and others will be found in the introduction and indexes of Leonardi's "I codici" (1959-60).

tianus' for eight hundred years—a revival of interest in the third generation of the Carolingian Age, and a flaring of interest in the eleventh century, culminating at Chartres (particularly in astronomy) in the twelfth century—but by the close of the Middle Ages the fortunes of the two authors diverged sharply. Macrobius' *Commentary on the Dream of Scipio*, presenting, as it does, the best Latin exposition of Neoplatonism and one of the finest specimens of Cicero's prose, was cherished during the Renaissance revival of interest in Platonism and Ciceronian style, and appeared in nearly fifty printed editions, including five separate incunabula editions.[5] Martianus' *De nuptiis*, generally discarded as a handbook of the seven liberal arts and having lost its popularity as a model of allegory, was not nearly as attractive as Macrobius to later ages and appeared in surprisingly few editions. These are listed here; they have been checked by the standard catalogues: for editions before 1501, by those of the British Museum, Bibliothèque Nationale, Hain, Panzer, Graesse, Polain, Pellechet, Guarnaschelli-Valenziani, and Goff; and for editions after 1500, by those of the British Museum, Bibliothèque Nationale, Graesse, and Brunet.[6]

Editions before 1501

1499 December 16, Vicenza, Henricus de Sancto Urso: BM XXXIII. 558; BN XXIII.501; Hain 4370; Panzer III.521; Graesse II.40; Polain 967; Pellechet 3224; Guarn.-Valenz. 2426; Goff 154.

[5] Macrobius (tr. Stahl), pp. 60-63.

[6] British Museum, *General Catalogue of Printed Books: Photolithographic Edition to 1555* (London, 1959-1966); Bibliothèque Nationale, *Catalogue général des livres imprimés* (Paris, 1897-); Ludwig Hain, *Reportorium bibliographicum . . . usque ad annum MD* (2 vols., Stuttgart and Paris, 1826-1838); Georg Wolfgang Panzer, *Annales typographici* (9 vols., Nuremberg, 1793-1801); Johann Georg Theodor Graesse, *Trésor des livres rares et précieux* (7 vols., Dresden, 1859-1869); M. Louis Polain, *Catalogue des livres imprimés au quinzième siècle des bibliothèques de Belgique* (4 vols., Brussels, 1932); M. Pellechet, *Catalogue général des incunables des bibliothèques publiques de France* (3 vols., Paris, 1905-1909); T. M. Guarneschelli and E. Valenziani, *Indice generale degli incunaboli delle biblioteche d'Italia* (3 vols., Rome, 1943-1965); Frederick R. Goff, *Incunabula in American Libraries: A Third Census of Fifteenth-Century Books Recorded in North American Collections* (New York, 1964); Jacques Charles Brunet, *Manuel du libraire et l'amateur de livres* (5 vols., Paris, 1860-1864).

1500 May 15, Modena, Dionysius Berthocus (reprint of 1499 ed.): BM XXXIII.558; Hain 4371; Panzer II.153; Graesse II.41; Polain 968; Pellechet 3225; Guarn.-Valenz. 2427.

Editions after 1500

1532 Basel, Henricus Petrus: BM XXXIII.558; BN XXIII.502; Graesse II.41.

1539 Lyons, Apud haeredes S. Vincentii: BN XXIII.502; Graesse II.41.

1577 Basel, P. Perna (with Isidore *Origines*): BM CXIII.7; BN XXIII. 502-3; Graesse II.41.

1592 Lyons, B. Vincentius (reprint of 1539 ed.): BM XXXIII.558; BN XXIII.502; Graesse II.41.

1599 Leiden, C. Raphelengius: BM XXXIII.559; BN XXIII.503; Graesse II.41; Brunet I.1558.

1619 Lyons, Apud haeredes S. Vincentii (reprint of 1539 ed.): BN XXIII.502.

1658 Lyons, J. A. Huguetan et M. A. Ravaud: BN XXIII.502; Graesse II.41.

Editions of parts of the work

1500 Erfurt, Wolfgangus Schenck: Hain 4372; Graesse II.41; Goff 154 (Book III).

1507 Frankfurt, Nic. Lamparter et Balth. Murrer: Graesse II.41 (Book III).

1509 Leipzig, Mart. Herbipol.: Graesse II.41 (Book V).

1516 May 11, Hieron. Vietor: BM XXXIII.559; BN XXIII.502; Graesse II.41 (Books I-II).

1763 Bern, Wagner: BM XXXIII.559; BN XXIII.502; Graesse II.41 (Books I-II).

1794 Nuremberg, Monathus et Kusslerus: BM XXXIII.559; BN XXIII. 503; Graesse II.41 (Books I-II).

Important collections containing single books

1652 *Antiquae musicae auctores septem Graece et Latine*, ed. M. Meibomius. Amsterdam, Apud L. Elzevirium: BM CLVII.135 (Book IX).

1863 *Rhetores Latini minores*, ed. C. Halm. Leipzig: BM XCVI.579 (Book V).

Translations (Books I-II only)

1578 *Le nozze di Mercurio e di Filologia*, tr. Alf. Buonacciuoli. Mantua, Fr. Osanna: Graesse II.41.

1629 *Le nozze dell'Eloquenza con Mercurio*, tr. Eureta Misoscolo. Fr. Pona. BN XXIII.503; Graesse II.41.

1837 Althochdeutsche . . . Übersetzung und Erläuterung [by Notker Labeo] der von Mart. Capella verfassten 2 Bücher *De nuptiis Mercurii et Philologiae*: BM XXXIII.559; BN XXIII.503.

The princeps edition by Franciscus Bodianus was published at Vicenza in 1499. Bodianus boasts in a prefatory letter that he corrected two thousand errors in the text.

The Vulcanius edition of Martianus (with Isidore), published at Basel in 1577, was noteworthy for including many of the Remigius glosses. A large number of these were introduced by Kopp into the footnotes of his 1836 edition.[7]

The outstanding edition before the nineteenth century, that produced by the sixteen-year-old prodigy Hugo Grotius, was published in Leiden in 1599.[8] The "Miracle of Holland," as Grotius was greeted by the French king when he went to Paris in 1598 on a diplomatic mission, was undoubtedly assisted in his editing by Joseph Justus Scaliger, his mentor at Leiden. The edition is marred by many rash emendations, but it is greatly enhanced by the ingenuity of its editor or editors and by its use of one of the best Martianus manuscripts: Leidensis 88. A copy of the Grotius edition in the British Museum contains in the margins emendations penned in the hand of Richard Bentley, England's greatest textual critic.[9] The irony involved in this col-

[7] See Remigius (ed. Lutz, I, 40) for the use of Remigius' glosses by other modern editors.

[8] Casaubon and Vossius both spoke with glowing admiration about Grotius' accomplishment in editing Martianus. See M. de Burigny's *Life of Hugo Grotius* (London, 1754), p. 15. Mark Pattison, the author of the "Grotius" article in the eleventh edition of the *Encyclopaedia Britannica* says: "In the annals of precocious genius there is no greater prodigy than Hugo Grotius, who was able to make good Latin verses at nine, was ripe for the university at twelve, and at fifteen edited the encyclopaedic work of Martianus Capella."

[9] A. Stachelscheid discusses these emendations in "Bentleys Emendationen von Marcianus Capella," *Rheinisches Museum*, XXXVI (1881), 157-58. (For a page of the Grotius edition showing Bentley's emendations, see above, the frontispiece.)

laboration must not pass without remark: we have here the text of a despised and "ignorant" writer, called by Scaliger "barbarus scriptor," being elucidated by three of the most prodigious intellects ever to engage in Latin textual criticism. The young Grotius may have been the last person to read *The Marriage* as it was intended to be read—for edification and amusement. He read Latin as if it were his vernacular and he was conversant with the technicalities of the ancient disciplines.

M. Meibom edited Book IX (*De harmonia*) in his edition of *Musicae Scriptores Antiqui* (Amsterdam, 1652). Meibom was an expert musicologist and he used two of the better manuscripts in his collation: Leidensis 36 and 88. His edition of Martianus' Book IX is still considered worthy of collation.

The voluminous edition by U. F. Kopp, prepared as a labor of love in his last years, was edited for the press by K. F. Herrmann at Frankfurt-am-Main in 1836. Kopp had lost his position as private secretary to the elector of Hesse and had turned to philology for consolation. His collation of fourteen manuscripts provided a broad base but he did not handle his manuscripts as expertly as later editors.[10] The chief value of Kopp's edition today is in his elaborate commentary. Until it is superseded by an up-to-date commentary of the same scope, Kopp's work will continue to be indispensable for Martianus scholars.

F. Eyssenhardt's Teubner edition (Leipzig, 1866) was prepared during a seven-year period when he also edited Macrobius, Ammianus Marcellinus, Apuleius, Phaedrus, and the *Historia miscella*. Eyssenhardt was able to produce so many texts in so short a period because he collated few manuscripts for his editions. For Martianus he used only two manuscripts, Reichenauensis and Bambergensis, and occasionally Darmstadtensis, when he found mutilations in the other two. Unimpressive as it appears, Eyssenhardt's edition should not be overlooked, because he was a scholar who read manuscripts with care and skill and he was thoroughly familiar with the literature and Latinity of the period.[11] Eyssenhardt's elaborate investigation of Martianus' sources in his introduction is still worthy of consideration.

[10] Herrmann remarks upon Kopp's editorial practices in his lengthy Preface to Kopp's edition, pp. ii-xx.

[11] His edition of Macrobius did not deserve the scathing strictures the eminent Wissowa expressed in his review in *Wochenschrift für klassische Philologie*, XII

Adolf Dick was occupied with editing Martianus throughout his professional career. He published his doctoral dissertation, *De Martiano Capella emendando*, in 1885, and his Teubner edition did not appear until 1925. He remarks in his preface about wartime delays and expresses regret that the publishers did not permit him to include in his Introduction his studies on the life, style, sources of the author, and his own emendations of the text.[12] Defects in Dick's edition have been pointed out above.[13] Dick's collation, though embracing fewer manuscripts than Kopp's, was a better one, and he handled his manuscripts with greater skill. In addition to the two main manuscripts used by Eyssenhardt (Dick was unable to examine the Darmstadt copy used by Eyssenhardt), Dick collated eleven other manuscripts, all on his own inspection. He was mistaken in supposing that none of the manuscripts in his collation was earlier than the tenth century—as Bischoff, Préaux, and Leonardi have noted.[14] Dick's text is the most reliable at present. It will be superseded by the edition that James Willis is preparing for the Teubner Library. Jean Préaux is at present engaged in completing an edition of Books I and II which will be based upon the best available manuscripts and will include a commentary embracing the glosses of Martin of Laon and of the two anonymous Cambridge commentaries referred to above,[15] as well as a French translation. This edition will appear shortly in the "Collection Budé."

(1895), 689: "Mein Urteil, dass wir es mit einer von Anfang bis zu Ende nachlässigen und unbrauchbaren Arbeit zu tun haben, glaube ich im vorstehenden ausreichend begründet zu haben."

[12] *Martianus* (ed. Dick), pp. xxv-xxvi. One can sympathize with the publishers, who had waited forty years for a manuscript and were undoubtedly despairing of its completion in Dick's lifetime.

[13] See pp. 62, 73-74. See also the reviews of Barwick in *Gnomon*, II (1926), 182-91; Manitius in *Philologische Wochenschrift*, XLV (1925), 543; Baehrens in *Jahresbericht über die Fortschritte der klassischen Altertumswissenschaft*, CCVIII (1926), 18-22; Lammert, *ibid.*, CCXXXI (1931), 75-77; Beeson in *CP*, XXI (1926), 92-93; and the comments and bibliography of Leonardi, "I codici" (1959), pp. 456-57.

[14] See Préaux, "Le Commentaire de Martin de Laon," p. 438; Leonardi, "Raterio e Marziano Capella," p. 82, n. 1.

[15] P. 64.

PART II

The Allegory and the Trivium

The Allegory and the Trivium

THE WORK of Martianus Capella may be considered as comprising three parts: the first two books of allegorical narrative, the trivium, and the quadrivium. Of these, the most important in the history of learning was the quadrivium, the principal subject of this volume. The allegory and the trivium of Martianus, though less important, nevertheless have had a significant role in intellectual history. As William Stahl has already pointed out, the allegorical framework of *The Marriage* exerted a considerable influence on medieval letters.[1] And the trivium books of Martianus were as important as his quadrivium books in establishing the seven liberal arts as the standard program of education in the Middle Ages.[2]

THE ALLEGORY AND ITS SOURCES

The setting of Martianus' work, in a council of the Olympian gods, one might think to be a mere literary convention, with a history extending back to Homer; but it is clear that to Martianus it is far more. His allegory teaches that the union of learning (Philology) and eloquence (Mercury) is a goal sanctioned by suprahuman authority, by all that is divine; that the curriculum of the seven liberal arts, being the means to achieve this goal, bears the same sanction; that through prowess in these studies and the benefits one thus brings to mankind, it is possible to win immortality and the fellowship of the gods.

Even apart from this allegory, there is in the first two books of the work a wealth of religious doctrine. Although the principal figures bear the names of Olympian gods, the religion is very different from that of Homer or Aeschylus. The setting is not Mount Olympus but the celestial spheres, and astral religion is fundamental to the thought. With this is blended Neopythagoreanism, old Roman and Etruscan

[1] See above, pp. 22-23, 25-26, 38, 55ff. See also Appendix A.

[2] See above, pp. 21-23.

religious ideas, some Neoplatonic concepts, some Egyptian deities, more than a trace of Hermeticism. The strands interweave into a whole, just as strands of vastly diverse thought interweave in modern Christianity; but in the one case as in the other, inconsistencies between strands may be discerned.

The serious allegory, the solemn religion are depicted in scenes of the most elaborate fancy and gaudy imagination, with much striving after and little achievement of real solemnity, and with frequent attempts at lightness of touch which are nothing more than inept frivolity. The discordance between the weighty messages Martianus feels impelled to express and the scenes, images, and language in which he chooses to express them makes his work a kind of sad classic in the history of didactic literature. This discordance is more readily understood, however, if one bears in mind Martianus' unhappy choice of models for his allegory.

As we have seen,[3] the principal model is the Cupid and Psyche episode from Apuleius' *Metamorphoses*. There is a mortal woman espoused to a god, a conclave of the gods to ratify the match, the apotheosis of the woman, her ascent to heaven. There, too, is an allegory, of the human soul (Psyche) falling in love with Love (Cupid) and enduring suffering before ultimate union. All of this is treated by Apuleius with a delicate touch, sometimes humorous, always lively and imaginative, that has ensured the tale's popularity as a romance from century to century. Martianus' debt to Apuleius appears time after time in general situations, detailed scenes, and innumerable echoes of words and phrases. There is no question that Martianus was inspired by Apuleius, not only with the idea of an allegorical wedding but also with the romantic and festive treatment proper to a wedding.

The gaiety borrowed from Apuleius accords ill with the tone of solemnity and profundity which also pervades the work as Martianus sets out religious doctrines after the manner of an inspired mystic. In this gravity—and in much of the material which is thus expounded—Martianus' work resembles Macrobius' commentary on Cicero's *Dream of Scipio*. The elements of astral religion, the sense of eternal destiny and divine sanction for human actions, and the didacticism about the

[3] See above, pp. 27, 32, 42.

physical universe are strongly similar in the two works. In Macrobius' commentary the tone is consistent with the theme of the work and is maintained throughout, just as Apuleius' very different tone is appropriate to his theme and consistently maintained. Martianus tried without success to harmonize the two approaches, so that we move from the levity of the first ten sections to an apocalyptic vision (11-22), back to the lighter vein, on to banality (at the end of 28), then wavering between solemnity (29-30), puerility (31), and solemnity again (32), until we reach a nadir of bad taste (at the end of 34 and 35). So it goes on.

In details of diction Martianus borrowed most heavily from Vergil and Ovid. This does not necessarily mean that he was especially fond of these poets. They were standard authors for literary study in the Roman schools and were popular sources of situation, phraseology, and illustration in the schools of rhetoric of the Empire, so that frequent borrowing from them is normal amongst late Latin writers. More significant are Martianus' clear echoes, on two occasions (sections 1 and 123), of Claudian; both, but especially the first, are relevant for dating *The Marriage* to the early fifth century.[4] While there are echoes of other authors, none are as frequently echoed as Vergil and Ovid. The almost total absence of Ciceronian echoes in Books I and II, together with the frequent misquotations and occasional misunderstandings of Cicero in Book V, lead one to wonder whether Martianus—though himself possibly a lawyer[5]—had any close acquaintance with the whole speeches (as distinct from extracts and from the technical rhetorical works) of the greatest of Roman legal pleaders.

RELIGIOUS IDEAS

The Marriage opens with a hymn to Hymen the god of marriage. Aside from the obvious dictates of tradition in such an opening, Hy-

[4] Martianus' treatment of Hymen the god of marriage (1) recalls Claudian's *Epithalamium of Palladius and Celerina* (*Carmina*, ed. Theodor Birt, *Monumenta Germaniae Historica*, 10 [Berlin, 1892], poem xxxi), lines 31-55; this poem probably dates from A.D. 399. Martianus' poem in section 123 recalls Claudian's *Panegyric on the Consulship of Flavius Manlius Theodorus* (*Carmina*, ed. Birt, poem xvii), lines 100-12; this consulship was in A.D. 399.

[5] See above, pp. 17-20.

men here has an allegorical role: he is presented as that divine concord, that principle of unification, which permeates and integrates a universe made up of infinitesimal elements—a physical concept which originated as early as Empedocles, and then passed into Platonic philosophy.[6] With this keynote Martianus establishes that his physics, and in particular his cosmology, is broadly Platonic. It is within the framework of a Neoplatonic explanation of the universe and of all being that Martianus' religious beliefs are set. This is not to say that he is in simple terms a Neoplatonist; many strands make up his religion. But his explanation of the universe, insofar as it is rational, is Neoplatonic,[7] and the religious ideas are brought into a measure of consistency with this.

Although the nine books, particularly Books I and II, are set in the heavens amongst the gods, there are three passages where the religious lore is especially rich. One describes the summoning of the inhabitants of heaven to a conclave (41-65); in the second Juno answers Philology's question about "what goes on in the vastness of the sky" (150-68); the third is Philology's hymn to Apollo (185-93). The first has been closely studied by Stefan Weinstock,[8] who corrected much of the earlier speculation by Carl Thulin;[9] the second has been discussed by Robert Turcan.[10]

From these passages it is clear that while the names and many characteristics of the Olympian deities have been retained, the Olympian pantheon has been integrated into the astral religion which permeates Neoplatonism, Neopythagoreanism, and Stoicism. Their abodes are around the zodiac, and they are identified with celestial bodies. The representation of Mercury varies from an anthropomorphic bridegroom (5, 35, and *passim*) and messenger of the gods, to the Neoplatonic Mind (92), to a planet with a fixed and known orbit (8, 25, 29). Apollo is at one time the god of prophecy sitting on a rock at

[6] Plato *Timaeus* 32c, *Gorgias* 508a; for Empedocles, see H. Diels and W. Kranz, eds., *Fragmente der Vorsokratiker*, I, 287-90, esp. passages 28, 29, 33, 37-40.

[7] See above, p. 10.

[8] "Martianus Capella and the Cosmic System of the Etruscans," *Journal of Roman Studies*, XXXVI (1946), 101-29.

[9] "Die Götter des Martianus Capella und der Bronzeleber von Piacenza," *Religionsgeschichtliche Versüche und Vorarbeiten*, Vol. III pt. 1 (1906).

[10] "Martianus Capella et Jamblique," *Revue des études latines*, XXXVI (1958), 235-54.

Delphi (10-11, 20-22, 26), at other times the sun (25, 29), at others the syncretization of a host of religious figures (191-92), and again the Neoplatonic Mind (193). This conversion of the Olympians into deities of other, philosophically based, religions was frequent in late antiquity.[11] In Martianus, however, not only the Olympians are thus absorbed, the sacred figures of ancient Italic belief are similarly treated: Lars lives in the second region of heaven, between Mars and Juno (46); Consus in the tenth region, with Neptune (54); Vejovis in the fifteenth, next to Saturn (59). What were at one time separate and relatively homogeneous currents of belief—philosophical, astrological, chthonic, anthropomorphic—have coalesced. The mixture is enriched by a large stream of numerology, which adds significance to any kind of group relatable to number. In all of this, particular attention is paid to Mercury, not only as the personification of eloquence but also as a god who enjoyed a special cult in North Africa.[12]

In this rich blend of faiths a reasonably consistent "theology" can be discerned. Throughout most of the work, Jupiter is presented as the supreme deity. It is his consent that must be obtained for the wedding, his word that convenes the council of the gods. He has displaced Saturn, the former ruler, now relegated to a minor role. So far Martianus accords with the traditional Olympian mythology. However, in section 185 and 202 we glimpse the Supreme Deity, the Unknown and Unknowable of Neoplatonic thought,[13] from whom, by the emanation of intermediaries, all being, including the array of Olympian gods, has its existence. This blend of Neoplatonism and Olympian religion does not necessarily involve a contradiction; it simply pushes the hierarchy of being one stage higher, postulates one or more deities of more exalted status, and fits the irrational Olympian figures into an intellectually defensible philosophic system. An infinite number of minor deities can then be accommodated, either by simple addition, just as the eighty-four attendants (200),[14] the innumerable genii and daemones

[11] See J. Bayet, *Histoire politique et psychologique de la religion romaine*, pp. 211-54.

[12] W. Deonna, *Mercure et le scorpion*, pp. 38-40.

[13] See esp. Plotinus *Enneads* 6. 9. 4-7.

[14] See Turcan, pp. 237-39.

(152-53) are introduced; or by syncretism of gods from faiths originally diverse, as Latin, Greek, and Asian religions meet in the wife of Saturn (3) or Egyptian, Libyan, Persian, and Phoenician gods are syncretized in Apollo (191-92).

The dwelling places of all these deities are distributed in the sky, both in plane and in elevation, as a draftsman might say. Their celestial abodes are described as lying in a plane around the 360 degrees of the zodiac (45-60); they are also described (150-66) as being at different levels in elevation, from the celestial sphere above the sun, where the most exalted gods live, to the regions between sun and moon and then between moon and earth, in a descending scale of being until we reach the surface of the earth, inhabited by mankind. Since men's souls are made of fire (originating from the divine stars) and have a natural tendency to rise if unhindered by the body, the purer souls may, when released from the body, ascend more or less high in this scale of being. Thus a concept akin to the Christian concept of salvation, of eternal felicity merited by life on earth, is implicit in this originally Neopythagorean system;[15] and related to this concept is the attitude of asceticism which condemns the body as impure, an obstacle and hindrance to the divine fire—an attitude rejected in theory by orthodox Christianity but nonetheless influential on much of medieval Christian thought. We begin to see another reason for the popularity of Martianus' work in monastic culture.

However, whereas the Christian attains salvation by faith, trust in God's mercy, and love manifested in deeds—none of which virtues demand intellectual gifts—the men who in Martianus' system attain immortality are (with the exception of Hercules) men whose wisdom, in matters of religious lore, agriculture, and technology, or the seven arts, has benefited mankind. The idea that an untutored peasant, by the mere quality of his love for God and His creatures, may attain sanctity and eternal bliss, is alien to Martianus; immortality in his eyes is earned by fame won through service, not by love or innocence alone. He was the last Latin exponent of what Marrou calls "the religion of culture,"[16] salvation through *paideia*. The wedding of Mercury and Philology allegorizes the union of eloquence, and intellectual prowess which

[15] F. Cumont, *After Life in Roman Paganism*, pp. 24-26.

[16] H.-I. Marrou, *A History of Education in Antiquity*, pp. 100-1.

makes that prowess effective and serviceable to mankind. May there not also be a further dimension to the allegory: that Mercury, who as Hermes Psychopompos conducts the souls of the elect after death to beatitude,[17] is the spouse who brings Philology, the learning of mortal men, to the eternal society of the gods?

In this striving after immortality men are aided or impeded by fate (11-15, 21-22, 32, 88), which implements the decisions of the gods (18, 64-65, 68-69); they are greatly assisted by intellectual power, which discovers the compulsions binding even on the gods (22); they may also obtain by propitiation the help of the gods; each individual has his personal guardian spirit (151, 160) and at the same time must contend with malevolent deities and spirits (47, 163-65). To find his way through the turmoil of this life, man should seek the will of the gods by all kinds of divination—through birds, thunder, entrails, prophets—and even through numerological calculation (893-94); he should also try to appease, propitiate, and influence the will of the gods by sacrifice and the tending of their shrines. There is no ethical teaching discernible in Martianus, no suggestion of moral laws or guidance for personal conduct comparable to the Mosaic commandments or the Christian beatitudes. There is also no suggestion of initiation into a mystery, of salvation through divine intercession, such as Apuleius' *Metamorphoses* reveals. Immortality comes not from divine gift or personal holiness but from the fame won by intellectual achievement, after effort and sacrifice.

Apart from this interpretation, the religious lore in *The Marriage* is of considerable interest to the historian of religion. The description of the assembly of the gods (41-65) is one of the three major sources for our pitifully slight knowledge of Etruscan religion,[18] which was so highly regarded by the Romans. *The Marriage* admirably exemplifies the confluence of religious traditions in the late Roman Empire and the role of decayed Neoplatonism in blending the streams. Both Weinstock and Turcan agree that this blending is not the work of Martianus, but that he used the now-lost teaching of Cornelius Labeo, who had brought late Greek and Italic religion and cosmology into a synthesis. Martianus' particular achievement was to combine that religious syn-

[17] Cumont, p. 25.

[18] Weinstock, p. 101.

thesis with the secular "religion of culture" by placing his treatise on the seven arts in that context.

THE CURRICULUM OF THE SEVEN LIBERAL ARTS

To a modern, the curriculum of the seven arts at first appears to have no principle of unity; it seems to be a random grouping of subjects in which any substitution of one for another would be no more significant than the substitution of geography for Spanish in a modern pupil's course of study. However, teachers of the early Middle Ages—Cassiodorus, Isidore, Alcuin, Rabanus Maurus, John of Salisbury, Thierry of Chartres[19]—regarded it as an integrated curriculum with seven components, all necessary. The origins of the curriculum were in classical Athens.

It is well known that the study of rhetoric began in the Greek world of the fifth century B.C. and was marketed in Athens by the Sophists;[20] Aristotle is reported to have regarded Zeno as the founder of dialectic;[21] grammar, the technical study of language, of the etymology and usage of words, also began with the Sophists.[22] The mathematical studies are older than this, but it was in classical Athens, especially in fourth-century Athens, that the two groups of studies—the literary-linguistic and the mathematical—came together to form a curriculum. The rhetorician Isocrates regarded mathematics (in moderation) as an acceptable propaedeutic to rhetorical study,[23] while Plato prescribed that the guardians of his republic study literature in their boyhood before they approach the mathematical and dialectical period of their training.[24] In the next generation, Aristotle expected pupils to have

[19] Cassiodorus *Institutiones II praef.* 1-2; Isidore *Etymologiae* 1. 2.; Alcuin *Grammatica* (Migne, *PL*, Vol. CI, cols. 853-54); cf. pseudo-Bede, *Elementorum Philosophiae libri* IV (Migne, *PL*, Vol. XC, col. 1178); Rabanus Maurus *De clericorum institutione* (Migne, *PL*, Vol. CVII, cols. 395-404); John of Salisbury *Metalogicon* 1. 12; 1. 24; 2. 9. On Thierry, see Édouard Jeauneau, "Le *Prologus in Eptatheucon* de Thierry de Chartres," *Mediaeval Studies*, XVI (1954), 174 f.

[20] See, e.g. G. Kennedy, *The Art of Persuasion in Greece*, chap. 2.

[21] See Sextus Empiricus *Adversus mathematicos* 7. 7; Diogenes Laertius *Lives of the Philosophers* 8. 57; 9. 25.

[22] This is the period in which Plato's *Cratylus* is set.

[23] *Antidosis* 264-69.

[24] *Republic* 376-98b.

studied literature dialectic, rhetoric, and mathematics before they came to him for philosophy.[25] The first extant writer whose works cover the essentials of the seven liberal arts is Heraclides of Pontus, a pupil of Aristotle who wrote on grammar, rhetoric, dialectic, music, and geometry, besides his philosophical works.[26] Even at this early stage, then, the people who were most interested in the full span of subjects were philosophers; and the seven liberal arts were in essence, and always remained, a philosophers' curriculum.

This may seem an exaggerated statement, considering that, as Marrou points out, the philosophers were not alone in fostering this program: Cicero and Quintilian, for example, considered the liberal arts to be the base of the ideal orator's education. In Marrou's words: "In the Roman epoch, the *encyclios paideia* appeared at least theoretically to be the necessary preparation for all forms of higher culture: literary, technical, scientific, as well as philosophical."[27]

Nevertheless, there is a world of difference between lip service and fulfillment. Quintilian, for instance, never shows any sign of proficiency in, or real concern with, the mathematical studies of the seven (geometry, arithmetic, astronomy, and music); Cicero, in his translations of Aratus and the *Timaeus*, shows more genuine interest—but then Cicero was more of a philosopher than Quintilian. The program of the liberal arts was no more than an unattainable ideal; "no longer the object of a regular education, it was merely a frame which each man's erudition strove to fill more or less."[28] The only people who seriously promoted the study of all seven liberal arts were philosophers, to whom alone the last four studies were important, for they are branches of the mathematical studies prescribed by Plato.[29]

The justifications for the first three studies in the curriculum were simple. Grammar covered both the elements of language—which we still call "grammar"—and the study of literature, especially poetry; it was thus the minimum introduction to one's cultural inheritance, the foundation of all education. Dialectic was a training in logic, a formal train-

[25] See Marrou, *Saint Augustin et la fin de la culture antique*, pp. 221-22.

[26] See Diogenes Laertius *Lives* 5. 86-88.

[27] Marrou, *Saint Augustin*, pp. 222-23 (my translation).

[28] *Ibid.*, p. 226.

[29] *Republic* 7. 525a-31e.

ing in verbal thinking; rhetoric, training in expression. They correspond to such subjects in modern English-speaking schools as English grammar, English literature, English expression, and logic—which may not all be taught as formal subjects in any one school today but are nevertheless part of its education.

So much for the three subjects which the Middle Ages called the trivium. The quadrivium, as Klinkenberg says,[30] was conceived by Boethius "as a genus whose species are the four mathematical disciplines of arithmetic, geometry, astronomy, and music. What gives it unity is the subject with which it deals: number, or rather magnitude. Arithmetic deals with magnitudes as such, geometry with immovable magnitudes, astronomy with magnitudes in motion, and music with the relations of different magnitudes to one another." The philosophers studied these subjects because, to quote Boethius, "everything that is formed from natural origins seems to be formed on a numerical basis. For this was the design foremost in the mind of the creator."[31] The science of number, arithmetic, is the key to the other three studies: geometry is the study of number given shape (we recall Martianus' description of Four as "the sure perfection of a solid body, for it comprises length breadth and depth" [734]); astronomy is the study of such shapes given motion (and furthermore, the stars in Platonism are divinities with special relationships to the souls of men); while music, the discipline of number in its proportions, was considered the key to all the relationships, physical and spiritual, quantitative and qualitative, of the world. According to Boethius, the cosmos is held together by number: "You bind the elements with numbers so that cold consorts with flames, dry things with liquids, so that the purer element of fire may not fly away or their weight drag down the submerged lands."[32] We are reminded instantly of the opening invocation of Martianus' work.

For probing the secrets not only of the physical world but of divinity and of the human soul as well, the quadrivium is an essential pre-

[30] Hans Klinkenberg, "Der Verfall des Quadriviums im frühen Mittelalter," in Josef Koch, ed., *Artes Liberales von der antiken Bildung zur Wissenschaft des Mittelalters*, 2 (my translation).

[31] Boethius, *De Arithmetica* 1. 2; Migne, *PL*, Vol. 63, col. 1083b.

[32] *Consolation of Philosophy* 3. m. 9. 10-12; cited by Klinkenberg, p. 2.

liminary. It serves two purposes: it trains the mind in the mathematical concepts and skills necessary to comprehend and investigate the mathematical basis of the world and the life of the world; and it purifies the soul by leading it to dwell on immaterial things, abstractions, and thus removes it from the life of nature to that of Soul and Mind. The latter purpose is well-known from Plato's exposition in his *Republic*, and was an argument which justified these studies to the Christian Clement of Alexandria.[33] Of course, mathematics had often been studied by scientists for its own sake, and the mathematical and astronomical works of Hipparchus, Eratosthenes, Euclid, and others had no religious motive. But in late antiquity, with the new impetus which Neoplatonism brought to philosophy, there was a new religious emphasis on the purification of the mind and heart. With the resurgence of a philosophy which used number, harmony, shape, and the stars as essentials in its ethics and metaphysics, these four mathematical studies reasserted their importance. Together with the literary studies they formed a total of seven, a number of great mystical significance —so much so that Augustine, who included philosophy in his list of liberal studies, omitted arithmetic (which would in any case be assumed as underlying geometry, astronomy, and music) in order to keep the magic number.[34]

In modern times we justify mathematical studies in our curricula either as a training in spatial, numerical, and nonverbal thinking or, pragmatically, as a preparation for a great many types of jobs and situations in everyday life. To the ancients the second consideration did not apply at the level of education we are discussing, for these were "liberal" studies, and were thought to be above the trivialities of earning a living. (This is hidden allegorically under Apollo's remark about Medicine and Architecture: they are not to speak at the wedding of Philology and Mercury because they are too occupied with mundane matters [891].) The first consideration was carried further by the ancients than we would carry it, to the point that they regarded these studies as a purification of the mind preparatory to mystical contemplation of truth. This justification, linked with their belief in the

[33] *Stromata* 6. 10-11.

[34] Marrou, *Saint Augustin*, p. 192. See above, p. 7, n. 11.

necessity of mathematics for understanding God, man, and the world, was obviously accepted only by philosophers and was dismissed by rhetoricians, who regarded these studies as impractical. The curriculum of the seven liberal arts, therefore, was fully taught only by philosophers, and widely accepted only when a resurgence of philosophy coincided with a decline of rhetoric. A sign of this coincidence may be read in the fact that between Varro, in the first century B.C., and Martianus, there is no evidence that any handbook of the seven liberal arts was written; yet, contemporary with Martianus, Augustine started one; in the next century, Cassiodorus compiled one, while Boethius wrote on many of the subjects, discussed the basis of the liberal arts, and appears to have coined the word "quadrivium" (or, in his form, *quadruvium*); and Isidore of Seville, at once an offspring of antiquity and a sire of the Middle Ages, treats of them in the first three books of his *Etymologies*.

Concerning the state of education in Martianus' time we learn almost nothing from Martianus himself; but, if our dating is right, he is almost an exact contemporary and fellow countryman of Augustine, whose education is well documented and has been exhaustively studied.[35] Augustine was brought up in the literary and rhetorical educational tradition and turned to philosophy only in maturity; he therefore provides an excellent picture of rhetorical education in its dotage.

The first study after learning to read and write was grammar, in its two senses of literature and linguistic structure. The linguistic structure taught in North Africa in the fourth century A.D. was the Latin of Rome of the first century B.C. It concentrated on morphology, deriving the rules of syntax more from the forms of inflected words than from their function in expressing meaning; by grammatical "errors," its teachers meant the deliberate irregularities and licenses found in some classical writers, especially poets—not the ignorant errors likely to appear in the Latin of a fourth-century provincial boy. The treatment of literature was if anything even more contrary to modern ideas: it consisted mainly in commenting on the text word by word, pointing out grammatical form and function, meaning (a fourth-century teacher might well have had to paraphrase a classical author much

[35] See Marrou, *Saint Augustin*.

as a modern teacher does with Shakespeare), rhetorical figures, etymology, any pertinent history or mythology (especially for proper names), and in general any item of information which the understanding of a particular word might require. Such a procedure could give the pupils a broad mass of historical, geographical, and other knowledge in the course of literary studies, and to that extent it provided a form of general education; but this knowledge was inevitably disorganized, derived haphazardly from single words as they occurred in a literary context. Moreover, this procedure could ruin the work as literature—never treating a passage as a whole, always atomizing it, breaking the continuity, emphasizing the trivial at the expense of the sublime.

Dialectic was often treated in antiquity as the counterpart of rhetoric. The Stoic Zeno had used the image of a hand:[36] the clenched fist is dialectic, compressed and forceful; the open palm is rhetoric, expansive and wide-ranging. The purpose of dialectic in the curriculum was to train the power of reasoning, to discover and fortify the arguments which rhetoric would then use. It was a subject of little interest to others than philosophers and its place in the trivium was effectively as the handmaid of rhetoric. Not until the twelfth century did it come into its glory.

The next major study after grammar, and for many the only other study, was rhetoric, generally according to the formal rules laid down by Cicero, with examples drawn from his speeches. The teaching of rhetoric in Martianus' time had not changed much since Quintilian: first the terms, divisions, and rules of rhetoric, then the elementary exercises, finally the *controversia* and *suasoria*, the declamations. The political themes were still drawn from the experiences of the Roman Republic or even the earlier Greek city-states, though Rome had been a monarchy for centuries; the legal themes were still those of Quintilian's day, though law had become increasingly a specialist's province. The main areas left to the orator as fields for his talents were display oratory and writing.

The narrowness of this education, so apparent to us as we list its sub-

[36] Cicero *Orator* 32; *De finibus* 2. 6. 17; Sextus Empiricus *Adversus mathematicos* 2. 7.

ject matter and procedures, escaped the notice of its teachers and most of its students because of the great attention it paid, after its fashion, to comprehensiveness. Did it not require the study of all the seven liberal arts as essential for any educated man? And if these studies were "taught" not as coherent intellectual disciplines but as scraps of information like those picked up at random in the grammar class—a procedure which to us vitiates their educational value—it does not mean they were dismissed as unimportant. "Learning" was desired and admired in a man insofar as it might help his oratory. Metaphors from astronomy, appeals based on ethical arguments, examples drawn from history (or mythology), these were sought and valued. So came those handbooks of many subjects which Martianus used as sources, pocket histories like that of Valerius Maximus; and though the "encyclopedias" had had a different genesis, they too were put to this use.

This tradition of polymathy, or universal learning, was an old-established one. Hippias the Sophist in the fifth century B.C. had an encyclopedic range of interests;[37] Aristotle had tried and must have practically succeeded in mastering the whole field of learning in his day; the scholars of Hellenistic Alexandria had tried by condensing and epitomizing to reduce the field of knowledge to manageable proportions. No one, however, seems to have tried alone to write an account of all that is known until Varro in the first century B.C. His works covered not only specialized treatises on the Latin language and on agriculture but also a survey of the nine arts (including medicine and architecture), a vast collection of essays in mixed verse and prose (the *Menippean Satires*), and a long work, the *Antiquities*, which was a primitive encyclopedia. Varro was emulated by Pliny the Elder, whose nephew has left us a vivid picture of his uncle's "scissors-and-paste" method: his slaves would read aloud to him the works of others, while he told them what passages to excerpt and copy out.[38] The consequences—secondhand information, an uncritical approach, inconsistencies, failure to acknowledge sources, lack of structure—are a foretaste of Martianus. Although many handbooks on individual subjects or groups of subjects were prepared in the interval, from some of

[37] Plato *Hippias Major* 285b-86c; Philostratus *Lives of the Sophists* 495.

[38] Pliny the Younger *Letters* 3. 5.

which Martianus did his own excerpting, there seems to have been no treatise with the range of Varro's or Pliny's; for when Augustine, at the same time as Martianus, felt impelled to write on the seven liberal arts, it was to Varro that he turned for a model. Martianus' work, derived from such sources, themselves derivative, could be produced only in an old and failing civilization: centuries of true scholarship, the whole range and depth of Greek and Roman learning, lie behind its stock definitions its trivialities and inconsistencies. "Are all thy conquests, glories, triumphs, spoils, shrunk to this little measure?"

Because *The Marriage* was later used as a school text, it has often been assumed to have been written as such. Yet its range is too wide and its treatment too inadequate, compared with the scope of the ancient school texts we know, for this to have been its genesis. Because it contains many Neoplatonic and Neopythagorean elements, it might appear to have been a contribution for the pagan opposition to Christianity; but at best this would apply only to the first two books; the last seven are ideologically neutral, and the work as a whole has no polemical or exegetical purpose at all. Because it appeared in a time of crisis and collapse, it has been thought to have been intended as a summary of the learning of antiquity, an "encyclopedia," to be transmitted to posterity before the barbarian invasions; but again the scope is too narrow and deliberately restricted. Furthermore, such a plan presumes a degree of foresight and a sense of foreboding of which Martianus gives no indication.

These elements are not to be disregarded: the seven liberal arts were at base a school curriculum; the work is that of a pagan using Neoplatonist terms; it did appear at a time of crisis. But these features are not sufficient to explain the work. In the words of Claudio Leonardi:

> The attitude, the taste, which here prevails and can explain it is not only a custom of erudition or a Neoplatonist faith; it is an attitude of "decadent" culture, and a manner of self-justification and defense by putting everything on display. It is a question of a human attitude and a cultural reaction clearly explainable in a moment of declared crisis or decadence: a defense by the parading of all one's "property", one's accumulation of learning.[39]

This purpose of display accounts for several otherwise unaccountable

[39] Claudio Leonardi, "Nota introduttiva per un'indagine sulla fortuna di Marziano Capella nel medioevo," *BISIAM*, LXVII (1955) 270, n. 67.

features of the work. For this reason, it covers all the seven liberal arts traditionally necessary to the formation of an educated man. For this reason, it employs the language and style of pedantic display, the tortuousness and bombast of a writer straining to impress his readers with his literary skills, and to show off his knowledge of Greek terms, Greek proper names, Greek declensions. For this reason, does a book on the seven arts have a gaudy mythological framework, laden and fretted with allegory, to our eyes so disproportionate in size and inappropriate to the main body of the work. Martianus is displaying his learning, in a religious context, perhaps to win for himself the immortality that Plato and Aristotle, Cicero and Varro, achieved. And, with all his faults, has he not been to some extent successful?

THE TRIVIUM

We have already observed the origins of rhetoric, dialectic, and grammar as three unrelated subjects, amongst the Greeks of the fifth and fourth centuries B.C. Of these, the subject with the most prestige came to be rhetoric; but rhetoric involves the use of words and the study of literature, as well as the use of logical argument, so that it is closely related to its sister arts, grammar and dialectic. The cementing of this relationship seems to have been primarily a Stoic contribution.[40] The Stoics divided the field of philosophy into ethics, natural philosophy, and logic; and this last included all aspects of verbal expression, hence the whole trivium. Sandys mentions Zeno himself and his successor in the school, Cleanthes, as authors of works on grammar[41] (Zeno's well-known comparison of dialectic and rhetoric has already been mentioned here). The later Stoics Chrysippus and Crates of Mallos certainly wrote and lectured on grammatical questions, and Crates introduced these studies to Rome.[42] In the first century B.C. the leading figures of Roman intellectual life, Varro and Cicero, who exerted the most profound influence on all subsequent Latin scholar-

[40] See especially Diogenes Laertius' appendix to his life of Zeno: *Lives* 7. 38-160, esp. 39-44, 55-60, 132-60.

[41] J. E. Sandys, *A History of Classical Scholarship*, Vol. I, pp. 148-49. See also G. Pire, *Stoicisme et pédagogie*, pts. 1 and 2.

[42] Suetonius *De grammaticis* 2.

ship in antiquity, were themselves pupils of the Stoics Aelius Stilo and Posidonius, respectively.[43] There were a Stoic approach to rhetoric and a Stoic theory of grammar, just as there was a Stoic logic; and, by their scholarly interest in these fields, combined with their studies of cosmology and their quasi-religious philosophy, the Stoics had the most important influence in establishing in the minds of the general educated public both the connection between the trivium and the quadrivium and the connection between the seven arts and philosophy.[44] In Varro's encyclopedia and Cicero's rhetorical works, grammar, dialectic, and rhetoric are the only literary-linguistic subjects which find a place in the program of studies. This reflects their accepted position in the *encyclios paideia*, in which they remain until Martianus' time.[45]

GRAMMAR

The study of grammar began with the Sophists of the fifth century B.C. and was carried forward by Plato[46] and to a lesser extent by Aristotle.[47] They may be passed over cursorily because the most rapid, wide-ranging, and influential development is found in the Stoic school, particularly (in this early period) in the work of Chrysippus. He is known to have written on the parts of speech, on cases and number, on solecisms, and especially on anomaly.[48] The Stoics, with their ethic of a life in accordance with nature and their intense concern with questions of natural philosophy, almost inevitably were committed to the view that language is a natural process in which there may be resemblances but not "rules" and in which coinages, variety, "anomalies" are to be expected. An opposing view was taken by the Alexandrian grammarians such as Aristophanes of Byzantium and Aristarchus and, later, Dionysius of Thrace; these men, with no philosophical commitment but a mission to preserve and restore the texts of classical Greek

[43] *Ibid.* 2-3; Cicero *Brutus* 205-6. See also Gudeman, "Grammatik," in Pauly-Wissowa, Vol. VII, pt. 2, cols. 1798-1800.

[44] See especially Pire, pt. 2, chap. 3, on Posidonius.

[45] For a fuller account of the development of the *encyclios paideia*, see Marrou, *History*, pp. 176-85; and *idem*, *Saint Augustin*, pp. 211-35.

[46] In the dialogue *Cratylus*.

[47] See Gudeman, cols. 1785-87.

[48] Diogenes Laertius *Lives* 7. 192.

literature, found assistance in determining the forms of archaic and unusual words by recourse to analogy with similar, more familiar words. The Alexandrians became the champions of "analogy," the grammatical theory that language is subject to rules, that words fall into certain classes and patterns of usage, and that there is little scope for variation within the classes. The inflected nature of Greek gives particular strength to this view, because in fact the range of terminations is extremely limited in comparison with the extraordinary richness of the vocabulary.

Thus in its early stages "grammar" was in many ways similar to the study now known as linguistics. Ancient grammar studied sounds (vowels semivowels, consonants, syllables in various aspects); it studied the formation of words and syllables and their changes; and it discussed the theories of language, anomaly and analogy. It was indeed far removed from modern introductions to Greek or Latin grammar, because it was not the study of a language—that is, the mastery of a tongue foreign to the student—but rather the study of language, with the scholars using their native tongue for purposes of illustration.

This study, together with the study of literature, which in antiquity was always included under the title of "Grammar," was introduced to Rome by Crates of Mallos about 169 B.C.[49] Since Crates was a Stoic, Stoic grammatical theories were understandably first in the field at Rome; the wave of Stoicism in ethics that also swelled in Rome in the second and first centuries B.C. naturally fostered the Stoic approach to grammar. Thus the first significant Roman grammarians, L. Aelius Stilo and Nigidius Figulus, were in the Stoic tradition. And Stilo's pupil was Varro, whose treatises on the liberal arts and on the Latin language provided source material for so much of the subsequent Latin grammatical writing. Varro, discussing at length both analogy and anomaly, took a somewhat inconsistent eclectic position.[50] However, in the first century A.D. a dissenter appeared, the manumitted slave Remmius Palaemon, whose massive *Ars Grammatica* (now lost) attacked Varro. By leaning heavily toward analogy, Palaemon clarified the Latin declensions and conjugations and clearly distinguished the

[49] See above, n. 42.

[50] See esp. *De lingua latina* 8. 23: "in my opinion we should follow both principles"; also 9. 3-6; 10. 60.

eight parts of speech. Subsequent Latin grammarians add little of significance to the work of Varro and Palaemon. The approach of the latter, which survives most fully in the treatise of Charisius, is evident in most later grammarians, including Martianus.

Martianus' grammar begins with a definition of the scope of the subject, then moves into an exhaustive discussion of letters, their possible positions in words, and their pronunciation. Syllables are next discussed, with considerable attention paid to pronunciation and accentuation. The discussion of final syllables is continued through each of the parts of speech, giving the reader some introduction to the declensions of nouns and the conjugations of verbs. These topics are considered at greater length in Martianus' next section, on analogy as applied first to nouns and adjectives, then to verbs. The treatise concludes with one section on anomaly, completely different in style from the rest of the treatise.

The sources of Martianus' Book III have been thoroughly investigated by Wilhelm Langbein,[51] following an earlier study by Johann Jürgensen.[52] The study is made particularly difficult, not to say inconclusive, because our surviving Latin grammatical treatises are mainly of the fourth century A.D. or later, and the most one can do is compare Martianus with each of these and construct hypotheses to explain similarities or differences.

There are many points of correspondence between Martianus' third book and the work of Diomedes (dating from the later fourth century A.D.), which Jürgensen explains by supposing that both borrowed from a common source. This may be true, but the source, if any, has perished. As Langbein points out, moreover, there are also many correspondences between Martianus and Charisius, Maximus Victorinus, Servius, and at least one anonymous author; and they cannot all be shown to descend from one common source. Furthermore, Martianus differs from Diomedes in his definition of a syllable, in distinguishing the three parts of a syllable, in distinguishing the four genera of *junctura*, and in details of the treatment of accent, including his assertion

[51] *De Martiano Capella grammatico.*

[52] "De tertio Martiani Capellae libro," *Commentationes philologae seminarii philologiae Lipsiensis*, 1874, pp. 57-96.

that a word can bear all three accents (e.g., Árgìlêtum). Langbein concludes that in these sections Martianus preferred to advance his own opinion.

There is an especially close similarity between Martianus and Charisius in their treatment of conjugations and declensions. Now Charisius derives from Cominianus, Julius Romanus, and ultimately Remmius Palaemon. Did Martianus follow Charisius or some one of these others? The resemblances are so close that Langbein concludes he followed Charisius directly; where he differs it is because he wished to advance his own opinion or perhaps because he followed a different source, such as Pliny the Elder. Martianus' treatment of words ending in *u* is not found in other extant grammarians. His treatment of the vowel in *lac* does not correspond with that of other grammarians though several discuss this point.

Martianus is close to Maximus Victorinus and Servius in his treatment of common syllables, final syllables, pronouns, verbs, adverbs, participles, conjunctions, prepositions, and interjections, but it is impossible to establish from which he drew. Though he has a few points of resemblance with Probus and with the Anonymous of Berne, these are slight compared to his correspondences with other grammarians. It seems clear that he and Priscian used the same source for description of the letters—whether the source is Papirianus or an earlier writer such as Probus or Pliny is again in doubt.

Langbein's conclusion is that Martianus' sources are of the fourth century rather than earlier grammarians; that he (as well as Charisius, Diomedes, Servius, and an anonymous author) borrowed from some sources now lost to us; that Martianus occasionally inserted into this material opinions of his own, of no particular merit; and that some of Varro's teaching, along with Palaemon's, found its way through intermediaries into Martianus' pages.

The later influence of this book appears to have been very slight. Gregory of Tours[53] mentions Martianus as a basic text in the seven arts, including grammar, but the reference is cursory and implies no special use of Martianus' Book III. Dick, in his edition of Martianus,[54]

[53] *History of the Franks* 10. 31.

[54] Pp. 106, 108, 110, 111, 113.

notes certain similarities between Book III and Bede's *De arte metrica.* One manuscript of Martin of Laon's commentary did not even bother to gloss Book III.[55] John Scot Eriugena's commentary[56] treats Book III very briefly, most attention being given to the allegorical introduction, and comments on the technical part of the book being confined almost entirely to explaining the uncommon words and proper names which occur amongst Martianus' examples. Remigius' treatment is even more striking,[57] for while he is normally much more full in comment than Eriugena, on Book III's technical material he makes no comment whatever; his notes deal exclusively with the allegorical sections.

Apart from manuscripts of the entire *Marriage*, twenty-three of the manuscripts listed by Leonardi[58] include Book III or some part of it. Two of these (nos. 76 and 219) are so slight as to be insignificant. Another (no. 122) contains the commentary of Remigius, but little of Martianus; another (no. 23) Leonardi describes simply as containing excerpts. Only two (nos. 77 and 189) contain Book III entire; these are an eleventh- or twelfth-century grammatical miscellany from Fulda (no. 77) and a ninth-century literary miscellany from Treviri (no. 189). The remaining seventeen manuscripts fall into two groups. Eleven of them (nos. 56, 85, 103, 104, 124, 127, 185, 192, 195, 201, 203) contain only section 261 of Martianus, on the pronunciation of each letter of the alphabet, and one (no. 138) contains sections 258-61. Of these, nine are dated from the ninth to the twelfth century, none from the thirteenth or fourteenth, three from the fifteenth and sixteenth. The second group contains Martianus' sections 300-9 (discussion of nouns ending in the letters *S* through *X*, and of voices of verbs) and 312-24 (conjugation of verbs) or portions of these. Of this group, four (nos. 74, 75, 149, 240) are dated between the eighth and tenth centuries, and one (no. 108) from the fifteenth. Three of the early manuscripts

[55] Bibl. nat. fonds lat. MS 12960, published by Cora E. Lutz, as *Dunchad: Glossae in Martianum.* The attribution to Martin of Laon and the identification of further manuscripts is by Jean Préaux, "Le Commentaire de Martin de Laon sur l'œuvre de Martianus Capella," *Latomus*, XII (1953), pp. 437-59. See above, pp. 4, n. 3, and 63-64.

[56] *Annotationes in Marcianum*, ed. Lutz.

[57] Remigius of Auxerre *Commentum in Martianum Capellam III-IX*, ed. Lutz.

[58] "I Codici di Marziano Capella," *Aevum*, XXXIII (1959), XXXIV (1960).

(nos. 74, 108, 149, 240) and the fifteenth-century copy associate the Martianus passages with Book II of Cassiodorus' *Institutiones* and other grammatical material.

The picture is fairly clear. Book III as a whole was almost unused. What value could it have had for people to whom Donatus' far superior grammar was available? The treatment of pronunciation was found useful and was often quoted; and a handbook made up of Cassiodorus' *Institutiones II*, Julius Severianus, a paschal computus, and excerpts from Martianus (300-9, 312-24) had a minor vogue.

DIALECTIC[59]

The subject of Martianus' fourth book is in most respects the same as the traditional formal logic derived from Aristotle which has been taught until recently in most university courses in "Logic." Now it has been increasingly superseded by "Symbolic logic," a more comprehensive study which includes Aristotelian logic merely as one of its subordinate parts. In Martianus' prologue (section 330), Dialectic herself claims to have originated in Egypt and to have emigrated to Athens with the philosopher Parmenides. Aristotle, however, is supposed to have attributed the discovery of dialectic to the philosopher of paradox, Zeno.[60] The alleged Egyptian origin may be put down to a persistent tendency to ascribe the finer products of Greek civilization to Eastern, and especially Egyptian, sources.[61] For instance Plato[62] in a myth attributes the discovery of number, calculation, and geometry (as well as astronomy, dice, draughts, and writing) to the Egyptian god Thoth, and in a symbolic sense he is probably right. However, the conception of arithmetic and geometry as systematic intellectual disciplines based on logical deduction seems to be purely Greek. In philosophy, Parmenides was the first to make conscious use of logical deduction as a way of seeking certain knowledge. Zeno advanced the subject by his use of the *reductio ad absurdum*, a procedure understood by Aristotle as dialectical, inasmuch as it involved positing the views

[59] This section is written by E. L. Burge.

[60] Diogenes Laertius *Lives* 8. 57; 9. 25.

[61] Kopp, in his edition of Martianus, cites Herodotus 2. 109; Diodorus 1. 69.

[62] *Phaedrus* 275c-d.

of an opponent and showing the contradictory consequences which they logically entailed.

In its origin dialectic was "a method of enquiry essentially involving conversation."[63] This is illustrated in the early dialogues of Plato by the conversational refutations to which Socrates devoted the greater part of his life.[64] The word itself does not occur earlier than the *Meno*[65] and may have been coined by Plato after reflection upon Socrates' philosophical methods. Throughout the later writings of Plato the word has a persistently favorable connotation, but its reference varies with Plato's changing views about which method promised to be most useful for investigating philosophical problems.[66] Socrates had used dialectical methods to convince men of their ignorance. In the *Meno* Plato holds out the possibility of achieving ethical knowledge by methods akin to those used with conspicuous success by the Pythagorean geometers. Dialectic now appears as a method of achieving positive doctrine, but retains from its Socratic ancestry an avowed passion for truth (in distinction from *eristic*, which sought merely to win arguments), a question-and-answer method, and a Zenonian concern with the logical consequences of a given position. By the *Republic* dialectic promised to achieve a higher form of knowledge than even mathematics.[67] Unlike mathematics, which works down from uncertified assumptions, it would be able to mount up to an "unhypothetized beginning" from which all knowledge could be deduced. As the highest point in the education of the guardians of Plato's ideal city, dialectic becomes "the coping-stone of all the sciences"[68] and synonymous with philosophy.

Dialectic's future, more humble position as a servant of rhetoric also had its origins in Plato. In the *Phaedrus* dialectic is indispensable to the

[63] I. M. Crombie, *An Examination of Plato's Doctrines*, I, 57, suggests that the active sense of the verb *dialegein*, "to discriminate," as well as the sense of the middle voice, "to converse," is contained in the derived term *dialectike*, and he therefore defines "dialectic" as "discrimination through cooperative discussion."

[64] *Apology* 21-23.

[65] Plato *Meno* 75d.

[66] R. Robinson, *Plato's Earlier Dialectic*, p. 70.

[67] Plato *Republic* 510-11.

[68] *Ibid.*, 534e.

true rhetorician.[69] Its role is threefold: to control the subject matter, occasion, and arrangement of discourse; to provide the "elevation of soul" necessary to complement the orator's natural endowments; and to study "the nature of soul," or, as we should say, psychology, in order to know what will persuade in any given case. The same passage introduces the equation of dialectic with the method of collection and division characteristic of the late dialogues, and influential in the formation of Aristotle's theories of classification, definition, and the syllogism. Despite the importance of his late dialogues and especially the *Sophist* to logic, Plato does not envisage the study of logic solely for its own sake,[70] or its later roles as either instrument or constitutive part of philosophy. An interest in logical and verbal puzzles for their own sake seems to have been the province of certain Sophists[71] and of a group founded in Megara by Euclides,[72] and known variously as Megarians, Eristics, and Dialecticians.

Aristotle shared none of Plato's optimism for achieving metaphysical or scientific truth by dialectical methods. He therefore distinguishes between philosophy and science on the one hand, and the principles of valid reasoning used in all disciplines employing argument and inference on the other. Being concerned with proof, Aristotle also distinguishes[73] between demonstration (valid argument from premises seen to be true) and dialectical reasoning (from premises adopted in debate but not necessarily true). His *Topics* are a handbook of dialectical arguments suitable for use in debate, the rules for which are laid down in the eighth book. His *Analytics* present the theory of the syllogism in the context of investigating demonstrative proof; but, as Aristotle himself acknowledges,[74] its principles are equally valuable for dialectical argument. By now, dialectic has come closer to what Plato would have termed "eristic," and its methodology involves a study of what the commentator Alexander of Aphrodisias (3d cent. A.D.) was later to call "logic." Henceforth the two separate strands of dialectic as the

[69] *Phaedrus* 269-74.
[70] W. and M. Kneale, *The Development of Logic*, p. 14.
[71] Aristotle *De Sophisticis Elenchis* 165a19-37; Plato *Euthydemus*, passim.
[72] Diogenes Laertius 2. 106.
[73] *Topics* 100a25-30; *Prior Analytics* 24a22-b12.
[74] *Prior Analytics* 24a26-28.

pure science of logic and the practical art of disputation become inextricably tangled.

As heir to this development, Martianus' compendium of dialectic appears excessively weighted with irrelevant logical material, if viewed as a debating manual, or sadly contaminated by its subservience to rhetoric, if viewed as the introduction to logic which it more nearly resembles. In addition, his work, in keeping with the long tradition of which it is part, includes a good deal of a more strictly philosophical character. This is because of the circumstance that Aristotle's early work the *Categories* was placed by his editors at the head of the more strictly logical works *De interpretatione*, *Prior and Posterior Analytics*, and *Topics* (of which the *Sophistical Refutations* originally formed part). As such it served as a general introduction to logic and philosophy in the Aristotelian tradition. In Martianus, sections 355-83 are based ultimately on the *Categories*. The preceding distinctions of genus, species, differentia, accident, proprium, and definition (344-49) are derived ultimately from the doctrine of the predicables in Aristotle's *Topics*.[75] Their position is due to Porphyry's influential *Introduction to the Categories*, written in the third century A.D. Martianus' discussion of how the single terms previously discussed are combined into complete sentences, and especially into assertoric propositions (388-96), has its distant ancestry in Aristotle's *De interpretatione*, and in Plato's *Sophist* before that.[76] The "square of opposition" (401-3) is ultimately derived from the same work of Aristotle, while the discussion of the conversion of propositions (397-403), like that of the three moods of categorical syllogism (404-13), is descended from the *Prior Analytics*. In this last-named work Aristotle had introduced letters of the alphabet to stand as variables in his inference patterns. In this way he achieved both clarity and generality, as well as showing that syllogistic reasoning was valid by virtue only of its formal characteristics. Martianus and his source reflect Aristotle's earlier practice of giving specific examples ("All justice is advantageous") which are to be understood in a general manner ("All A is B"). That Martianus shows no awareness of Aristotle's modal logic is not surprising. On the other hand his failure to treat of fallacies, which would be highly relevant to disputa-

[75] Aristotle *Topics* 1. 5-9.

[76] *Sophist* 261-63.

tion, results from a lack of proportion and planning in the rest of the book. The deficiency is ill concealed by Athena's allegations (423) that the teaching of sophism and deceptions is unpleasing to Jove.

With the exception of a few seminal references to the "hypothetical syllogism" in the *Analytics*,[77] Aristotelian logic is concerned with analyzed propositions, and the variables employed stand for single terms ("man," "justice," "advantageous," and so on). The passing suggestions of the *Analytics* were developed by Aristotle's successor as head of the Lyceum, Theophrastus, but the full development of a logic of whole propositions ("Justice is advantageous," "It is day," and so on) was the work of the Stoics. To mark the difference between their propositional variables and the term variables of the Peripatetic tradition they used ordinal numbers ("the first," "the second," and so on) in their inference schemata.[78] The use of this symbolism by Martianus (420) indicates a Stoic origin for his discussion of the hypothetical (*condicionalis*) syllogism (414-20).

Apart from this, and the reference to the greatest Stoic logician, Chrysippus, in the introductory allegory (327), there are other indications of Stoic influence in Book IV. One is the formation of the contradictory of a given proposition by prefixing the negative particle to the whole proposition,[79] and not merely to its verb (402). Offensive as this often was to idiomatic purists, the logical propriety of saying "Not: it is day" rather than "It is not day" can be seen by considering that this form alone shows unambiguously the extent of what is being negated, and by comparing with the modern symbolism "not-p." Again, the necessary use of the exclusive[80] form of disjunction (*aut*) in Martianus' fourth conditional mode (4-17) is characteristically Stoic.

Martianus' inclusion of arguments as "conditional syllogisms" which have a complex major premise but lack any "if" is due ultimately to Theophrastus. Reflecting on hints in Aristotle's logical works, Theo-

[77] Aristotle *Prior Analytics* 53b-12, 57b6. It should also be noted that Aristotle regularly sets out the syllogism as a conditional sentence, to the effect that if the premises hold so does the conclusion. The propositional *modus ponens* is thus presupposed by his procedure.

[78] Apuleius *Peri hermeneias* 279-80; ed. Thomas, p. 193.

[79] Sextus Empiricus, *Adversus mathematicos* 8. 89 ff., Apuleius *Peri hermeneias* 267; ed. Thomas, p. 177. B. Mates, *Stoic Logic*, p. 31.

[80] Mates, p. 51.

phrastus appears to have begun investigating arguments "from hypothesis"[81] in which a desired conclusion is agreed to follow, provided some other proposition (still to be investigated) is true.[82] This led to a recognition in the Peripatetic school of several arguments in the propositional form later investigated by the Stoics, and the extension of "hypothetical" to cover all arguments with a leading premise formed from more than one simple proposition. Martianus' presentation of the subject, in which the premises are drawn up with a view to a predetermined conclusion, is in keeping with this development, and with the Aristotelian view of the subject's dialectical, rather than logical or scientific, importance. It can thus be seen that Martianus is heir to a tradition which cared little for purity of allegiance to a particular school, Stoic or Peripatetic, but took what it pleased from either.

In the Stoic tradition, dating from its founder, Zeno of Citium, dialectic became the general name for the study of logic and other related topics, such as grammar and epistemology. As such it was with physics and ethics a constituent part of the total province of philosophy.[83] The choice of name shows the important connection of Stoic logic with the Megarian school mentioned above. The logicians of this school, preeminently Eubulides, Diodorus Cronus, and Philo, in succession to the Eleatic Zeno, had maintained a persistent interest in paradox, as well as in modal arguments and conditional propositions. Discussions of modalities (necessity, possibility, and so on), whether Aristotelian, Megarian, or Stoic, leave no trace in Martianus. Of the seven paradoxes attributed to Eubulides[84] two are referred to in Martianus' opening allegory (327), but are there associated with Chrysippus and the Stoic school. These are the "sorites,"[85] or "Heap" ("Would

[81] Aristotle *Prior Analytics* 40b25, 45b15, 50a32.

[82] See Kneale, pp. 98, 105. Our knowledge is due to Alexander of Aphrodisias' commentary on the *Analytics*.

[83] This threefold division is foreshadowed in Aristotle's classification of problems into ethical, physical, and logical (*Topics* 105b20).

[84] Diogenes Laertius 2. 108.

[85] "Sorites" is usually used by modern logicians to refer to syllogistic-type arguments with three or more premises. It is used in Martianus' concluding verses of Book IV (423) to refer to an accumulation of arguments which become imperceptibly more and more misleading as they proceed. See Cicero *Academica* 2. 49.

you say that this single grain is a heap?—No.—These two grains?—No.—These three? . . . etc. Then where do you draw the line?", and the "Horned Man"[86] ("What you have not lost you still have. But you have not lost horns. So you still have horns"). It should be noted that an interest in the analysis of paradox is a sign of the true logician, far from the trivial game it seems to those who lack understanding of the subject. It is therefore not surprising that Chrysippus should have concerned himself with the Megarian paradoxes, or that Martianus' understanding cannot go beyond making a witticism.

Yet another aspect of Stoic logic which leaves a trace in Martianus' terminology, if not his full understanding, is the distinction made between verbal signs and the meanings (*lekta*) which they signify. The latter were regarded by the Stoics as incorporeal and common to different languages. The class of *lekta* of special concern to the adherents of a propositional logic were those signified by complete sentences used to make assertions which must be either true or false. In Greek these were termed *axiomata*: the term was rendered into Latin as *enuntiatum* by Cicero and as *proloquium* by Varro.[87] Martianus' use of the term in section 389 and elsewhere, substantiates the suggestion of the prologue (335) that he is following a Varronian source. In Martianus, however, the use of the term is coupled with a reversion to an Aristotelian position where the *proloquium* itself (rather than the sentence which expresses it) is made of a noun plus a verb, and may be "doubtful" as well as true or false (390). This well illustrates M. Kneale's view that "for some centuries after Stoic logic had been formulated by Chrysippus we find discussion of the merits of his system and that of Aristotle, then a gradual fusion, or perhaps we should say confusion, which was completed at the end of classical antiquity in the work of Boethius."[88]

If the ultimate sources for the material in Book IV are clearly discernible, the channels through which it made its way to Martianus' pages are far less evident. Inevitably the name of Varro is the first to come to mind. The most striking Varronian characteristic is, as we have seen, the use of the term *proloquium*, which corresponds in most

[86] Diogenes Laertius 8. 187.

[87] Apuleius *Peri hermeneias* 265; ed. Thomas, p. 176, line 15. Aulus Gellius 16. 8.

[88] Kneale, p. 177.

respects to our "proposition." The use of *propositio* with this meaning is found in the *Peri hermeneias* attributed to Apuleius,[89] but not in Martianus, where it is reserved for the leading (complex) premise of a hypothetical syllogism (414) and introduced without definition or explanation. Cicero's use[90] of the term (and also the associated term *assumptio* for the minor premise) with the same meaning as Martianus suggests that it may go back to his teacher L. Aelius Stilo, who was also the teacher of Varro.[91] Stilo's book, rare in the second century, was clearly unknown to Martianus, and Martianus' discussion of the hypothetical syllogism is therefore likely to be Varronian. Whether Varro is being used directly it is impossible to say. It is noteworthy, however, that all but the third of the seven modes of the hypothetical syllogism given by Martianus correspond with the list in Cicero's *Topica*,[92] and that the one discrepancy can be attributed to Cicero's misunderstanding of the differences between the five basic Stoic inference schemata and the list of seven given in some other manual.[93] In the sixth century Cassiodorus excerpted a similar list of seven hypothetical syllogisms from a work by Marius Victorinus.[94] The differences are sufficiently great to exclude Victorinus as Martianus' source. Cassiodorus mentions an extended work by the otherwise unknown Marcellus of Carthage of which two books were devoted to Stoic propositional logic and were followed by a book on "mixed" syllogisms. The use of so full a treatment by Martianus seems unlikely, but cannot be ruled out, in view of his including a largely incomprehensible passage on mixed syllogisms (422), no doubt intended to give the suggestion of the greater profundities which could be dealt with if only Mercury and Athena would allow more time.

The Varronian term *proloquium* belongs to a series of related terms where the root is kept constant and the meaning altered by varying

[89] Apuleius *Peri hermeneias* 266; ed. Thomas, p. 177.

[90] *De inventione* 1. 57 ff. See also *Rhetorica ad Herennium* 2. 28.

[91] Aulus Gellius 16. 8. Stilo's book was apparently called *De proloquiis*, which suggests that the term *proloquium* originated with him.

[92] Cicero *Topics* 56-57.

[93] Kneale, pp. 179 81.

[94] Cassiodorus *Institutiones* 2. 13: in Migne, *PL*, Vol. LXX, col. 1173; ed. Mynors, p. 119.

the prefix. This method of coining technical terms is illustrated in the fragment of Book XXIV of Varro's *De lingua Latina* quoted by Aulus Gellius (16. 8), listing the kinds of complex propositions:

> Conditional proposition: *adiunctum* (or *conexum*)
> Conjunctive proposition: *coniunctum* (or *copulatum*)
> Disjunctive proposition: *disiunctum.*

The other terms of the series to which *proloquium* belongs determine the structure of Martianus' whole book, and are listed at the outset (338) as the first four of six canonical divisions (*normae*) of dialectic. Thus the discussion of uncombined terms (the Aristotelian predicables and categories) is called *de loquendo*; the discussion of terms combined to form complete sentences is called *de eloquendo*, the discussion of the subclass of these which make assertions and are thus either true or false is called *de proloquendo*, and finally the drawing of inferences from combinations of these is called the *summa proloquiorum*, "summation of propositions." The same terminology and principle of ordering the subject[95] is found in the fourth chapter of *De dialectica* by one Augustinus, which is included as an appendix to the works of St. Augustine. A comparative study of this work and that of Martianus by Fischer[96] shows that neither is a source for the other but that both rely on a third, either Varro himself or a close follower.

It would seem then that Martianus took a basically Varronian framework and expanded it with material from both his Varronian source and elsewhere. Comparison with the Augustinian document indicates that two passages are Varronian: First, in the passage on the combination of single terms (388-92), which corresponds (as noted by Kopp) to the Aristotelian *De interpretatione*, Martianus, as Dick observes, "follows other footsteps." The Augustinian document and Martianus here correspond closely in content and terminology, but not in detailed syntax or the examples given. It would be reasonable to suppose that Martianus is excerpting, while Augustinus is paraphrasing. An important difference is the use of the term *verbum*: in Martianus it cor-

[95] See Migne, *PL*, Vol. XXXII, col. 1410.

[96] B. Fischer, *De Augustini disciplinarum libro qui est de dialectica.* I have at the time of writing been unable to read Fischer's dissertation, and have therefore had to rely on brief notes taken some years ago by a colleague.

responds only to our "verb,"[97] in Augustinus it also corresponds to the Stoic *lekton*, whether in the sense of a single term or a complete utterance. The inclusion of an etymological discussion of *verbum* in Augustinus, coupled with his rejection of etymologizing[98] suggests not only that he had a Varronian work before him but that he was treating it critically. Second, the passage (355-60) which stands as an introduction to the ten categories (in place of Aristotle's explanation of what is meant by "homonymous," "synonymous," and "paronymous") also has affinities with the logical works included in the Augustinian corpus. The influential terms *aequivocum* and *univocum* do not seem to occur in extant authors before their occurrence in fourth-century Africa;[99] the term *plurivocum* is unique to Martianus. If Martianus coined it himself, Dialectic's apology for the word (339) is strange in the writing of Latin's most prodigious user of *hapax legomena*. It corresponds to the Greek term *polyonyma*, used in the same context at the opening of pseudo-Augustine's *Ten Categories*, where it is illustrated by the same example.[100] The discussion of the formation of words *per similitudinem*, *per propinquitatem*, and *per contrarium* (360), with the example of *lucus* derived from *lucere*, again has its counterpart in the Augustinian discussion of etymology.[101]

In the Latin West during the Middle Ages, the standard introduction to philosophy was provided by Boethius' translation with commentary of an *Isagoge*, or *Introduction*, to Aristotle *Categories*, written in Greek by the third-century Neoplatonist commentator Porphyry. Based upon Aristotle's discussions in the *Topics*, this work dealt with the predicables genus, species, differentia, proprium, and accident, and is important both for determining the order of presenting the subject and for inspiring the great nominalist-versus-realist controversy in the twelfth century. Boethius also wrote a commentary on the text of a fourth-century translation of Porphyry by Marius Victorinus. Examination of this shows that Victorinus' translation is rather free, sub-

[97] *De nuptiis* 388 ff. Cf. Varro *De lingua latina* 8. 11-13.

[98] Migne, *PL*, Vol. XXXII, col. 1412: *unde sit dictum non curemus, cum quod significet intellegamus.*

[99] *Ibid.*, col. 1416.

[100] *Ibid.*, col. 1421.

[101] *Ibid.*, cols. 1412-13.

stituting Roman proper names for Greek, on occasion summarizing,[102] and sometimes misinterpreting, the original text.

The same description could be given of Martianus' accounts (344-49, 361-87), which correspond to Porphyry's *Isagoge* and Aristotle's *Categories.* The immediate sources for this section are unknown, but no one who compares Martianus with the Greek originals can believe that he is handling them directly. His version of the *Categories* is several times differently ordered from Aristotle's both in its major divisions and in detail, and while in part it reads like an abridged translation, a great deal is in the form of an explanatory commentary. Some of this seems disproportionately elaborated in a presentation which otherwise gives only the technical terms and the barest outline of the subject. Of the predicables, the definition of "accident" is most puzzling; it is contrary to Porphyry's and to any other that I know. Martianus defines accident as qualities peculiar to one species. Ordinarily, white or black are taken as accidents of man, but clearly they apply to other species as well. Although it is tempting to suspect Victorinus as Martianus' main Aristotelian source, this is only a guess. Latin definitions of "genus" and "species," as well as "partition," are at least as old as Cicero's *De inventione* (I. 32), and would have been available to Martianus from many sources, including his own general knowledge.

There remains Martianus' account of the quantity and quality of categorical propositions, conversion, the square of opposition, and the nineteen valid moods of the categorical syllogism (396-413). It is generally agreed that, like the similar account in Cassiodorus, that of Martianus is taken from the brief manual *Peri hermeneias*, attributed to Apuleius. The correspondence in context, language, and examples is very close. The main difference is that Apuleius, while aware of the term *proloquium* (and a number of other equivalents), prefers to use *propositio.* He is also concerned to contrast Aristotelian logic with Stoic, to the advantage of the former. This polemic leaves no trace in Martianus.

Even this, our most confident attribution of source, must, however, remain subject to doubt. This is because there are passages in Martianus

[102] See Busse, ed., *Commentaria in Aristotelem Graeca*, Vol. IV, pt. I, p. xxxi: *Victorinum non tam versionem Isagoges composuisse apparet, quam in compendii formam eam redegisse.*

which are expansions of the material in Apuleius,[103] and other passages[104] where Apuleius expresses the thought not only more clearly than Martianus but also more briefly. The possibility of a common Aristotelian source cannot be totally excluded. If, however, Apuleius is truly Martianus' prime source, one can only wish that Martianus had plagiarized more wholeheartedly, since the copy in no way improves upon the presumed original.

The influence of Martianus' Book IV emerges clearly from a study of Leonardi's census. Apart from the entire *De nuptiis*, fifteen of the manuscripts he lists contain all or part of Book IV.[105] Two of these (nos. 76, 205) are fragmentary, and one other (no. 23) consists of notes from Book IV. The remaining twelve contain either the entire book or the entire technical discipline, omitting the allegorical setting. Thirteen manuscripts are dated between the ninth and twelfth centuries, with the biggest concentration (five manuscripts) in the tenth century; two are dated between the fourteenth and sixteenth centuries. Two manuscripts (nos. 135 and 136) came from Fleury in the ninth or tenth century; one (no. 151) from western France and one (no. 205) from Orléans at the same time; one (no. 172) from Bec in the twelfth century. It would seem that this book enjoyed a vogue in northwestern France up to the end of the eleventh century but was little used after the dramatic development of logic from that time onward.

RHETORIC

The study of rhetoric originated in Sicily in the fifth century B.C.[106] and was brought by the Sophists to Athens, where it found a ready audience in a political democracy with a passion for litigation. In the next generation, Isocrates was its principal exponent, Plato the leading

[103] For example, the Ciceronian example in section 399 has no counterpart in Apuleius.

[104] Cf., for instance, the expressions of the principle of "minimal interpretation": *quoniam id potissimum enumerandum, quod securum habet intellectum, indefinitum pro particulari accipitur* (Martianus 396) and *pro particulari semper valet, quia tutius est id ex incerto accipere, quod minus est* (Apuleius *Peri hermeneias* 266; ed. Thomas, p. 177).

[105] Nos. 18, 19, 23, 29, 76, 102, 123, 128, 131, 135, 136, 151, 172, 205, 239.

[106] See, e.g., Kennedy, pp. 26, 58-62.

voice of the opposition; a conflict between philosophy and rhetoric endured in Greek higher education for another two centuries.[107] Despite this conflict, Aristotle was the author of the earliest major rhetorical treatise we possess: His *Rhetoric* is a study of the principles implicit in the practice he observed in fourth-century Athens, and under the weight of his authority and the precedent of his analysis, the observations and classifications appropriate to the practice of Athens in his time became enshrined as timeless principles of rhetorical theory.[108] Aristotle's successor in the Lyceum, Theophrastus, contributed to the theory of literary style, but otherwise there was little advance in rhetorical theory to match the wide dissemination of its practice in education before the time of Hermagoras in the middle of the second century B.C.

Hermagoras of Temnos wrote on forensic oratory, in particular, on *inventio*, the discovery of arguments. He laid great stress on the determination of the *status*, the essential issue in any given case, and formulated precepts for its discovery and treatment. Thus the emphasis in rhetorical teaching, which had for some time been on diction and style (generating the "Asianist" school, and later, by reaction, the "Atticist"), now became focused on the more objective and rational aspect, *inventio*. It was at this point of development that Greek rhetoric first began to command the serious attention of the Romans.

The earliest Roman treatise on rhetoric we possess, the anonymous *Rhetorica ad Herennium*, is probably of the early first century B.C.[109] It is therefore close in time to Cicero's first treatise, *De inventione*, and in late antiquity and the Middle Ages the two works together were generally regarded as Cicero's, despite the differences between them. Cicero's other major rhetorical treatises—*De oratore*, *Orator*, and *De optimo genere oratorum* on the ideal orator; *Brutus* on great orators of the past; *Topics* on argumentation; and *Oratoriae partitiones*

[107] See Marrou, *History* pp. 210-12.

[108] E.g., the division of rhetoric into three (and only three) types: bouleutic, forensic, epideictic; the division of the duties of the audience (*Rhetoric* 1. 3. 1-3). For the topical, or specifically Athenian, bias of the *Rhetoric*, see, e.g., 1. 1. 4-5, 10.

[109] *Rhetorica ad Herennium*, ed. Caplan, p. vii, but see A. E. Douglas, "Clausulae in the *Rhetorica ad Herennium* as Evidence of Its Date," *Classical Quarterly*, n.s. X (1960), pp. 65-78, which argues for a date of about 50 B.C. if not later.

on the division of rhetoric—generally received much less attention in later Latin scholarship before the Renaissance, despite their merits, which eclipse *De inventione*. Cicero's contribution to the development of rhetoric was threefold: first, as the author of the treatises mentioned, he transmitted a great deal of Greek theory to the Latin world; second, under the influence of his Stoic and Rhodian teachers he denied and tried to heal the breach between rhetoric and philosophy; third, as the outstanding Roman orator, he provided an inexhaustible repertoire of examples of all aspects of rhetoric in practice.

Cicero's later years coincided with the beginning in Rome of that practice of declamation which originated as rhetorical training and became virtually an art form in its own right, with a profound influence on Latin literature.[110] While this was an important development, it had little effect on theory, which in the first century A.D. was more occupied with the conflict between Apollodorus and Theodorus and their respective followers. The Apollodoreans had a relatively rigid, doctrinaire approach to the composition of speeches, which they regarded as a science reducible to firm rules; whereas the Theodoreans treated it as an art, flexible and adaptable to the needs of the moment in the particular case. As we shall see, some Theodorean influence appears in Martianus.

The last major Roman writer on rhetoric was Quintilian, at the end of the first century A.D. Quintilian's synthesis of ancient thought on this subject sets out in measured pace and at considerable length, lucidly and with abundant illustration, the way to produce orators to match the immortal Cicero. As a treatise on rhetoric it is unmatched for fullness, clarity, and intelligence by anything else surviving from antiquity. It was too good for Quintilian's successors, and did not really win the appreciation it deserves until the Renaissance.[111]

The first and second centuries A.D. saw in the Greek world the efflorescence of rhetoric known as the "Second Sophistic," with an accompanying spate of rhetorical textbooks of generally little value. The major exponent was Hermogenes, and he appears to have had no

[110] See S. F. Bonner, *Roman Declamation*, pp. 149-67.

[111] For a survey of its influence, and its limited popularity in late antiquity, see F. H. Colson, ed., *M. Fabii Quintiliani Institutionis Oratoriae Liber I*, pp. xliii-lxxxix.

influence on Martianus. At the same time and after, appeared a number of Latin handbooks on rhetoric, many of which have been collected by Halm.[112] In general they bear the same relationship to the major authors as do handbooks in the other ancient subjects to the primary sources: "in them we find a grand concoction of Cicero and Quintilian with the subtleties, often imperfectly understood, of the first- and second-century Greeks, and sometimes, too, with strains more directly descended from Hermagoras."[113] It is on men such as these that Martianus drew.

The two principal studies of the sources of Martianus' Book V are those of Hinks[114] and Fischer.[115] From Hinks' study, the more penetrating and precise of the two, are taken the following conclusions: Cicero's *De inventione* was a prime source for Martianus, both as a direct source of theory and as offering a framework for a treatise on rhetoric. Martianus seems to make frequent direct use of the *De inventione* in his earlier chapters when treating the definitions and the divisions of rhetoric. He takes from it some elements of the theory of constitutions and the subdivisions of *qualitas*. And finally he uses it for his treatment of exordium, proposition and partition. Martianus also uses Cicero's *De oratore* and *Orator* especially for the treatment of expression, memory, and delivery. On literary style he borrows a little from Donatus and minor grammarians and a long passage on figures (section 523-37) from Aquila Romanus.

Besides these fairly obvious sources are some less easy to uncover. Hinks[116] shows Martianus' use of commentaries by Marius Victorinus on Cicero's *De inventione* and *Topics*. These, while adding little but confusion to Martianus' theory, give us firm evidence for dating his work; for Victorinus, after his conversion to Christianity, was affected by the edict of the Emperor Julian in 362 forbidding Christians to teach in the schools, and in his last years Victorinus wrote theological

[112] *Rhetores Latini Minores*.

[113] D. A. G. Hinks, *Martianus Capella on Rhetoric*, pp. 4-5. This unpublished dissertation, in the library of Trinity College, Cambridge, is a brilliant unraveling of the sources of Martianus' Book V.

[114] See the preceding note.

[115] H. W. Fischer, *Untersuchungen über die Quellen der Rhetorik des Martianus Capella*.

[116] Pp. 38-41, 63-65.

works. Victorinus' commentaries on Cicero, in Hinks's words, "can scarcely be placed earlier than about 325 and may well be much later."[117]

For his theory of constitutions Martianus draws on three sources, often incompatible: one Ciceronian, one working on principles which we know from Quintilian (3. 2. 26) to be Theodorean and not found elsewhere in Latin, and one clearly Hermagorean. The Theodorean is unidentifiable. The Hermagorean is shown to be very likely Marcomannus, the source common to Martianus, Fortunatianus, and much of Sulpitius Victor. Nothing of Marcomannus survives, but Hinks with striking scholarship establishes as probable that he was a Hermagorean and an older contemporary of Hermogenes; that he wrote a commentary on *De inventione*, applying to it later Greek rhetorical theory; and that he was the source who mediated to Martianus Quintilian's teaching on memory and delivery as well as the work of an unknown theorist on *ductus causae* and some of the parts of speech.

Behind Martianus' sections 447-48 (on the *genera causarum*) there appears to be a lost source, probably Greek, with a better understanding of Aristotle's theory of this subject than is found in surviving Latin rhetoricians. This may have been transmitted through some other source, just as the unidentified Theodorean's material may have been.

It would appear that Martianus used a wider range of sources for his discourse on rhetoric than for most of his other handbooks, and attempted, with neither success nor consciousness of failure, to combine conflicting theories. Virtually all the sources have been laid bare by Hinks with an acumen no less admirable than his modesty. His final remarks are in more than one way illuminating. He explains that he has said nothing of Varro—although Varro is reasonably thought to be behind the third and fourth books (in the fifth book there seems to be no trace of Varro)—"because, if we saw a trace of Varro, we could not recognise it. Nothing whatever seems to be known about that book of his encyclopaedia which dealt with rhetoric; and where nothing is known I have forborne to conjecture."[118]

[117] Hinks, pp. 1-2.
[118] Hinks, pp. 122-23.

The later use of Martianus' Book V is very peculiar. Twenty of the manuscripts listed by Leonardi[119] contain some portion of Book V. In only two (nos. 39 and 61) is the entire book contained. Two others (nos. 239 and 241) contain the entire technical treatise, omitting the allegory; both of these are humanist manuscripts dated between the fourteenth and sixteenth centuries. Of the total of twenty manuscripts, thirteen[120] come from this period; the other seven, between the tenth and twelfth centuries. Of the thirteen, ten contain sections 508-25 (the introduction to *elocutio*, all the figures of thought, and the treatment of *clausulae*); most of these appear to come from northern Italy. Three others (no. 227, from the tenth century; and nos. 116 and 232 from the twelfth) contain Martianus' treatment of the figures of speech (531-37), with various additional material.

What is the complete picture? Book V was used sporadically, especially in northern France and Germany, in the tenth and eleventh centuries, somewhat more often in the twelfth; then, in composite rhetorical manuscripts of North Italian humanists, Martianus' treatment of *elocutio*, the figures of thought, and the *clausulae* were found especially interesting and were often reproduced. This accords well with the revived interest in Latin style which was characteristic of humanism; whereas for the full treatment of rhetoric all the wealth of Cicero and Quintilian was at hand.

Each of the three parts of *The Marriage*—the two books of allegory, the trivium, the quadrivium—had a life and influence of its own in European learning. But for a full appreciation the work deserves to be seen as a whole; and to be seen not as a collection of seven handbooks but as one philosophicoreligious work. The interrelations of the religious beliefs and numerology with the seven arts and with the literary style of the work can then be seen to be remarkably strong and complex. These interrelations, treated fully and discerningly by Fanny Le Moine,[121] cannot be extensively treated here; we have been

[119] Nos. 16, 22, 23, 39, 41, 54, 57, 61, 116, 156, 172, 184, 200, 213, 216, 217, 227, 232, 239, 241.

[120] Nos. 22, 23, 41, 54, 57, 61, 184, 200, 213, 216, 217, 239, 241.

[121] In a Ph. D. dissertation at Bryn Mawr College entitled *A Literary Re-evaluation of the De nuptiis Philologiae et Mercurii of Martianus Capella.*

able to examine only briefly the basic religious ideas and the union between them and the seven treatises. The weaknesses of Martianus' work, in content and in style, have been pointed out abundantly, yet when one speculates on the intent of the work as a whole, one may well concur with Miss Le Moine that it is a "grand attempt" and as such "deserves to stand in the long tradition of works which are consciously designed to present a synthesis of the total pattern of the cosmos."[122]

[122] *Ibid.*, p. 297.

PART III

The Quadrivium

On Geometry

BEFORE the bridesmaid Geometry makes her entrance into the celestial hall, two distinguished-looking attendants[1] appear, bearing an *abacus*[2] sprinkled with greenish powder. This object, it is explained, is designed for delineating geometric figures and can also represent the circles of the world, great and small. Geometry enters, carrying a *radius*[3] in her right hand and a globe in her left. The globe is a replica of the universe, wrought by Archimedes' hand.[4] In it

> The planetary orbs gleam in the dusk of night,
> As precious gems sparkle in a setting of gold.

[1] Satire a moment later identifies them as Philosophy and Paedia. These allegorical ladies remind us of two others, with the same names, who figured in the careers of prominent writers. Lucian, in his autobiographical dream (*Somnium* 9), introduces Paedia, the cultivated lady who drew him into a career of learning and philosophy. Boethius' *De consolatione philosophiae* is a dialogue between the author and an allegorical young-old female named Philosophy. Supernatural, young-old figures like these were frequently introduced into stories in antiquity and by the fifth century had become rhetorical clichés, according to Curtius, pp. 101-5. Curtius (p. 104) points to six of Martianus' bridesmaids as such allegorical stereotypes. I fail to see why he omitted Harmony, who, though not "described in detail," as he says, is depicted by Martianus (909) as a "lofty figure whose melodious head was adorned with ornaments of glittering gold" and as a maiden who "moved with a grace that her mother could hardly match."

[2] *Abacus:* a tray, covered with sand or powder, used by mathematicians for drawing diagrams.

[3] A geometer's rule (*virga geometricalis*), according to Remigius (ed. Lutz, II, 130). In medieval art the *radius* was also taken to be compasses, and in one instance Geometry is found with a rule in one hand and compasses in the other. See Émile Mâle, *Religious Art in France, XIII Century*, pp. 78, 85.

[4] Archimedes devised and constructed an orrery in which the relative motions of the planets and the celestial sphere were accurately reproduced and in which lunar eclipses could be observed. He regarded this sphere as his outstanding technical achievement. It was brought to Rome after Marcellus' capture of Syracuse in 212 B.C. For references to it in the Latin writers see J. G. Frazer's note to Ovid *Fasti* 6. 277 in his edition, Vol. IV (London, 1929), p. 204. According to

The peplos she wears is emblazoned with figures depicting celestial orbits and spheres; the earth's shadow reaches into the sky, giving a purplish hue to the golden orbs of the sun and moon; there are gnomons of sundials and figures showing intervals, weights,[5] and measures. Her hair is beatifully groomed, but her feet are covered with grime and her shoes are worn to shreds from treading across the entire surface of the earth. She can, Geometry says, describe any portion of the earth's surface from memory and knows the exact distance between the earth and the celestial sphere, down to the inch. As she steps forward to draw diagrams on the abacus, she notices Archimedes and Euclid among the wedding guests. She could call upon them to expound the discipline, but the occasion calls for rhetorical skill;[6] since they speak no Latin, Geometry herself will reveal the secrets of the art in Latin, a rare occurrence for this subject, she avers.[7]

Amidst the resplendent setting of the celestial canopy and the garment and accouterments of the lady herself, it is not the abacus and geometer's rule or the figures on her garment that hold the clue to the contents of this book; rather it is the tattered boots. Geometry reminds her audience of the literal meaning ("Earth-measuring") of her name. The bulk of the lengthy discourse that follows (the longest book in the entire work) is devoted not to definitions, axioms, and proposi-

---

Lactantius (*Divinae institutiones* 2. 5), the sphere was made of bronze; according to Claudian (*Carmina minora* 51), of glass. Perhaps the celestial sphere was of glass, the interior orbs of bronze. See also E. J. Dijksterhuis, *Archimedes* (Copenhagen, 1956), pp. 23-25.

[5] This is the only reference I have found in Martianus to a function of Geometry that was to become prominent in the Middle Ages and is named as her function in the medieval mnemonic for the seven liberal arts: *Gram loquitur, Dia vera docet, Rhe verba colorat, Mus canit, Ar numerat,* Geo ponderat, *As colit astra.* On the attributes of Geometry in medieval art and literature see Klibansky, Panofsky, and Saxl, pp. 327-38.

[6] We have here a fine example of the deleterious effect that the rhetorician's attitude had upon Roman science. Cicero (*De oratore* 1. 3-6, 11-16) expresses admiration for the accomplishments of Greek scientists and technologists, but he has higher regard for a Roman rhetorician's skill in expounding the results of a Greek theoretician's concepts and constructs. The Latin rhetorician did not make the effort to comprehend Greek scientific theory and was interested only in summarizing and applying its results.

[7] For the setting described here, see sections 575-88 of the *De nuptiis.*

tions but to a conspectus of the known world.[8] At the conclusion of her survey she admits that she has been digressing and that the topic she is about to take up is the true subject of the discipline (*artis praecepta*). Because the day is waning, however, she is advised to restrict herself to essentials, and to touch lightly on them (*summa quaeque praestringens*), so as not to vex her listeners.[9]

We naturally wonder why Geometry decided to present her bridal offering in the form of a geographical compendium instead of devoting her main attention to the appropriate quadrivium subject. Our perplexity might be relieved if we had certain knowledge of how Varro treated geometry in the fourth book of his lost *Nine Books of the Disciplines*. If our earlier surmises are correct,[10] that Varro's book comprised a modicum of geometry and the elements of surveying, Martianus must have been largely independent of Varro in composing his *De geometria*. Martianus' book makes no mention of surveying, and it is clear that Varro's book did not deal with regional geography.

We have some scant knowledge about Latin studies of geometry in the period just before and just after the fall of the Western Empire. All quadrivium studies languished in the Latin world, and geometry was the most neglected subject in the four.[11] Augustine, who, prior to Martianus, had been engaged in compiling manuals on the seven disciplines, abandoned the project after completing a *De grammatica* and a draft of six books (the *rhythmus* portion) of a *De musica*.[12] How

[8] Sections 590-702 are devoted to geography; sections 706-24 are devoted to geometry.

[9] 703, 705. It is standard practice among Latin compilers to offer some excuse for not dealing with the intricacies of Greek theoretical subjects, the usual excuse being that the subject would bore readers.

[10] See above, pp. 44-47.

[11] Cassiodorus (*Variae* 3. 52. 7) laments about the neglect of quadrivium studies in his day: "No one attends a lecture on arithmetic; geometry is a subject for specialists; . . . disputes about boundaries are left to professional surveyors." See Ullman, "Geometry in the Mediaeval Quadrivium," p. 264, on geometry as the private domain of *agrimensores*. Cassiodorus himself devotes thirty-eight pages to the trivium, twenty-six to the quadrivium, and only two to geometry (*Institutiones II*, ed. Mynors); Isidore devotes a book to grammar, a book to rhetoric, and a book to the entire quadrivium, in which geometry is given only three pages (*Etymologiae*, ed. Lindsay).

[12] See above, p. 7.

he would have treated geometry is not known. Boethius, who came after Martianus and was the first writer to use the term *quadrivium*, did complete a *De geometria* as part of his set of quadrivium manuals.[13] This work was a translation of all or part of Euclid's *Elements*. As often as medieval Latin fragments of geometry turn up and are found to be accurate translations of Euclid, they are generally considered by scholars to be vestiges of Boethius' version.[14] But, as we shall see, there are some fairly faithful translations of Euclidean propositions (without the proofs) in Martianus' seventh book, which show that good Latin translations of at least portions of the *Elements* were in existence before Boethius' time.[15] Cassiodorus touches ever so lightly on the discipline of geometry (*Institutiones* 2. 6) and his treatment could have contributed nothing to contemporary knowledge of the subject.

Why did Martianus prefer geography to geometry as a subject for his *De geometria* and why did he introduce more Euclidean material in Book VII than in Book VI? Each of his bridesmaids is an impressively erudite lady and each holds forth in a lengthy discourse. Martianus could not, as Cassiodorus did, give a two-page, discursive treatment to geometry. He had to find some subject to fill out Book VI to normal length, and one that would not repel his readers. The Euclid-

[13] Of Boethius' quadrivium the *De institutione arithmetica* and part of the *De institutione musica* survive (the last eleven chapters are lost); the *De geometria* survives in fragments, and the book on astronomy, though lost, is well attested to. For a text of the first two works and some of the geometry fragments, see the edition of G. Friedlein; for a discussion of Boethius' quadrivium books, see Stahl, *Roman Science*, pp. 198-201.

[14] George D. Goldat's *The Early Medieval Traditions of Euclid's Elements* is a valuable study of the extant remains of Euclidean geometry up to the middle of the twelfth century and of modern scholarship on this little-known subject. The Verona fragments (Goldat, p. 24) have just been edited by Mario Geymonat, *Euclidis Latine facti Fragmenta Veronensia* (Milan, 1966). Geymonat dates the fragments in the fifth century and believes that they belong to Boethius' version. If this thesis is correct, it would indicate, as Ullman points out (p. 272), that Boethius was a very young man when he composed his *De geometria*.

[15] Unfortunately more attention has been drawn to Martianus' geometry in Book VI than in Book VII, even though Book VII contains the greater amount of Euclidean material. See e.g. Goldat, p. 28; J. L. Heiberg, *Litterargeschichtliche Studien über Euklid* (Leipzig, 1882), pp. 202-3.

ean material in Book VII is on numbers, not geometry, and originated in the arithmetical books of the *Elements* (VII-IX). Geometry proper was obviously a subject that Martianus could not handle, and he did not suppose that his readers would be interested in it. His choice of subject—a geographical survey of the known world—appears to have been a unique one for a manual of the traditional disciplines. The reason for his choosing geography probably lies in the enormous popularity of the works which were his two chief sources, the *Collectanea rerum memorabilium* of Solinus, and Pliny's *Natural History*.

It is common knowledge that Pliny's encyclopedia was one of the most popular and authoritative works on Latin science throughout the Middle Ages; this is attested to by the frequent occurrence of his name in medieval library catalogues and by the abundance of Plinian excerpts in codices of scientific writings.[16] But the great bulkiness of his work militated against its use in entirety. Generally speaking, writers like Bede, who quote extended passages from Pliny verbatim, used only certain books and did not have access to the complete work. Solinus hit upon a clever scheme for reducing Pliny: he depended upon an existing epitome of the geographical books (III-VI) to construct his framework and introduced a few hundred choice tidbits from Pliny's nongeographical books (mainly from VII-XII, on man, zoology, and exotic trees; and from XXXVII, on precious stones) to produce a compilation which he frankly admits in his Preface is designed to catch the reader's interest. More than three-fourths of Solinus' *Collectanea rerum memorabilium* was ultimately derived from Pliny. Mommsen found approximately 1,150 extracts from Pliny and 38 from Mela in Solinus' book, which numbers less than one hundred pages.[17] Solinus' book of marvels—in all probability extracted from an early second-century epitome of Pliny's geographical books[18]—became

[16] Many of these are included in passing by Leonardi in his census.

[17] Theodor Mommsen, perhaps the greatest of modern classical scholars, lavished his expert attention on the preparation of his masterly edition of this trivial work. His list (pp. 238-49) of passages derived by Solinus from Pliny and Mela, and of passages derived by later authors (Ammianus Marcellinus, Augustine, Martianus, Priscian, Isidore, Aldhelm, Bede, Dicuil, and the author of the anonymous *De situ orbis*) from Solinus, is a valuable index to the deterioration of classical writings at the hands of medieval compilers.

[18] See above, p. 47.

even more popular than its well-known forebear. And, as we have seen, Martianus further digested the work of Solinus and Pliny to produce the geographical excursus which forms the bulk of Book VI.

The geographical writings of the Greeks reflect two salient cultural characteristics—a penchant for theoretical studies, including mathematics, and a love for, and dependence upon, the sea. In mathematical geography they were outstanding. The accomplishments of Eratosthenes, Hipparchus, Ptolemy, and others rank high in the history of geodesy.[19] Descriptive geography, on the other hand, was not a well-developed or well-defined field. From the beginning the distinctions between history and geography were blurred; historians like Herodotus regularly introduced geographical excursuses, and geographers felt impelled to deal with historical backgrounds.[20] Greek treatises on geography tended from earliest times to assume the form of a *periegesis*, a survey of cities, peoples, and countries arranged in the order that a navigator would come upon them as he sailed the coast of the Mediterranean and outer seas. The commonest starting point for such surveys was the Strait of Gibraltar.[21] A *periegesis* naturally gave disproportionate attention to coastal regions, to the neglect of the interior.

Geographical writings of the classical period are now represented only by fragments or by borrowings in secondary sources. The literature surviving from the Hellenistic Age is written almost exclusively on a popular level, and has a standardized form. Introductory chapters contain stereotyped doctrines of mathematical geography, such as are found in the handbooks of Theon of Smyrna, Geminus, and Cleomedes:[22] the celestial sphere and circles are defined, and the positions

[19] For a discussion of the work of these mathematical geographers, see Thomson, chaps. V and XI; and E. H. Bunbury, *A History of Ancient Geography*, chaps. XVI, XVII, and XXVIII.

[20] F. W. Walbank has written a brilliant article with fine insights into the popular traditions of Greek geography: "The Geography of Polybius," *Classica et mediaevalia*, IX (1947), 155-82. He attributes the geographical digression as a normal feature of historiography to the methods of the story-teller.

[21] Lionel Pearson, *Early Ionian Historians* (Oxford, 1939), p. 30, thinks that Hecataeus' lost *Periegesis* (*c.* 500 B.C.), the earliest Greek geographical treatise, may have begun its survey at Gibraltar.

[22] Theon, ed. Hiller, pp. 120-38; Geminus, Chaps. IV, V, XV, XVI; Cleomedes *De motu circulari* I. 1-3.

of planetary orbits and the earth are located; then follow discussions of the location of the earth's zones of human habitation and the division of the known world into three continents. Polybius interrupts the narrative of his *Histories*, just before Hannibal's crossing of the Alps, to introduce a digression (3. 36-38) on the three continental divisions. Strabo, after a historical introduction, devotes fifteen of the seventeen books of his *Geography* to a description of the countries of the world. But nestled between his historical introduction and regional description is a lengthy chapter (2. 5) containing all the stereotypes of a popular handbook on cosmography: the shape and the zones of heaven and earth, the relative position and size of the earth, the four zones of human habitation, a coastal survey of the known world, and lastly a discussion of the climates and the hours of daylight at each of the parallels at the time of the solstice. A similar introduction to cosmography is found in the pseudo-Aristotelian *De mundo*.

The Romans made their first extensive contacts with the Greek intellectual world in the second century B.C. Being unable to comprehend the theoretical treatises of the thinkers like Eratosthenes and Hipparchus, they depended upon popularizers for their technical information. Writers of the first centuries B.C. and A.D. gave a canonical form to Latin geographical treatises, dividing them into two parts: a concise account of the elements of mathematical geography; and a chorography of the known world, preponderantly of coastal regions and consisting largely of lists of place names, with remarks about associated persons or things interspersed. Since the Romans had little or no interest in geometry and their conquests and administrative operations were conducted in the interior more than in coastal regions, it is clear that Roman geographers blundered in adhering to traditional patterns of Greek geography. Formal Latin chorographies were of little use to provincial administrators and military commanders, who depended instead on the data compiled by Marcus Agrippa in a gromatic survey of the highways and provinces of the Empire, conducted under the sponsorship of Augustus. The results of this survey were made available in itineraries and copies of the master map, one of which, the "Peutinger Table," survives in a late form.

Mela offers the faintest traces of mathematical geography in a few sentences in the opening chapter of his *De situ orbis* (*c.* A.D. 42). He then moves on to the divisions of the world and to a chorography, by provinces, starting at Gibraltar. Pliny begins his monumental *Natural History*, thirty-seven books long, with a book on the universe (Book II), the conventional opening of Greek mathematical geographers, but he does not comprehend the authorities he is using. He then divides the known world into two inhabited islands, as it were, bisected by the Mediterranean and the Black Sea, and proceeds on a coastal survey of the northern half (III-IV) and the southern half (V-VI), each time starting at Gibraltar. Pliny shows a high regard for the data compiled by Agrippa and makes frequent use of them in recording distances; but he fails to realize that Agrippa's survey would have made a sound basis for his chorography, as a Greek-style *periegesis* (probably Varro's) did not. Pliny gives meager treatment to interior regions which were of vital importance to the Empire. Though he served in military campaigns in Germany and wrote an authoritative history of the German wars in twenty books, he devotes about one hundred and fifty words to this part of the world. The reason is clear. He is drawing from books rather than from his own experience.[23] Solinus omits a mathematical introduction altogether and begins his chorography with Italy, though he uses Pliny and Mela throughout as his sources.

Dividing the known world into halves and assuming that an ocean flowed continuously about both halves satisfied Greek geographers that a coastal survey would serve as an adequate basis for world geography. But to Roman geographers, aware of military campaigning and explorations in Britain, Germany, Upper Egypt, and the Middle East, it appeared that proofs were required to demonstrate the adequacy of coastal surveys to encompass remote regions. Latin writers customarily introduced proofs of a continuous ocean—fabulous circumnavigations of the northern and southern continents. Pliny's proofs are garbled by Martianus. Pliny (2. 170), on the authority of Cornelius Nepos, says that Metellus Celer received as a present from the king of the Swabians some Indians who had been driven from their course, by storms, all the way to Germany; Martianus (621) has it that Cornelius, after taking some Indians captive, sailed past Germany. Mar-

[23] He had stated earlier (2. 117-18) that writers can often gather more reliable data about their own region from books written by authors who have never been there than from natives.

tianus also repeats Pliny's fabulous reports of Augustus sailing past Jutland into the frozen Scythian Ocean, of Macedonian sailors cruising from the Indian Ocean to the Caspian Sea, and of shipwrecked hulls of Spanish ships found in the Arabian Gulf.[24]

Martianus' lengthy excursus on geography, being derived from Pliny and Solinus, follows the canonical form of Latin chorographies. The stock features of mathematical geography at the opening—proofs of the earth's spherical shape and its location at the center of the universe (590-601), an explanation of the terrestrial zones and four human habitations of the earth, and a listing of the dimensions of the known world (602-16)—come from the Pliny-derived chorography and generally follow Pliny's order.[25] The main part, the regional geography, follows Pliny's and not Solinus' order, beginning the surveys of the northern and southern halves of the known world at Gibraltar (617-702). The precise borrowings of Martianus from Pliny and Solinus have been traced by Lüdecke.[26]

Pliny's four books of geography, together with excerpts from other books, which had been reduced by Solinus to a treatise of less than

[24] On Roman corruptions of Greek reports, see Bunbury, II, 172.

[25] The phrase *sicut Secundus* (590), read by some scholars as part of the text, but considered by Dick (292. 19-20) to be a gloss, is an acknowledgment, either by Martianus or by the glossator, that the material to follow comes from Pliny (Plinius Secundus). That an intermediary, and not Pliny, was used by Martianus is evident from the considerable amount of garbling and misreading of Pliny's text. E.g., Pliny's phrase *serius nobis illi* (2. 180) becomes a nonexistent reporter, Servius Nobilis, in Martianus (594); Pliny (2. 178) has two appearances of the Bear at Meroë, in the evening at the summer solstice and at daybreak just before the rising of Arcturus; Martianus (593) combines the two: at the summer solstice, at dawn, about the time of the rising of Arcturus. On the stereotyped character of Pliny's and Martianus' elements of mathematical geography and for some classical parallels see Macrobius *Commentary*, tr. Stahl: p. 154, the earth occupies a point in space; pp. 181-82, the earth is at the center of the universe; the center is the middle and the middle is the bottom of the universe; p. 204, antipodes have similar climates; p. 206, there are four regions of human habitation, diametrically and transversely opposed to each other. See also below, p. 176. Isidore (*Etymologiae* 3. 30-46) repeats the elements of basic cosmography and Jacques Fontaine has devoted a lengthy chapter to them in his *Isidore de Séville*, II, 469-501. His elaborate documentation provides a mass of parallels in classical and medieval writers.

[26] See above, "Sources," n. 27.

one hundred pages, were condensed by Martianus into an excursus one fourth as long as Solinus' book. Pliny's thousands of place names diminish to hundreds in Martianus. If Pliny lacked good sense, he did not lack industry. He gives distances between places and between boundaries en route. Many of these figures are omitted in Martianus' text, and totals are sometimes incorrect. Solinus culls from Pliny the better-known place names and those with entertaining associations. When Martianus further distills Solinus the result is ludicrous. Whole paragraphs are compressed into tortuous sentences, such as this horrendous example (668): "Not far distant are seven mountains which, because of their equal height, are called 'The Brothers'; they teem with elephants and lie beyond the province of Tingitana, whose length is 170 miles." Excisions inevitably lead to confusion, as in the section on Ceylon. Pliny (6. 84-88) reports the visit of Ceylonese ambassadors to Rome and their surprise at finding that the sun in northern latitudes rises on the left (of an observer looking south toward the sun); Martianus (697) omits the report of the ambassadors' visit to Rome and has the sun rising on the left in Ceylon. Pliny (6. 82) says that the sea between India and Ceylon is generally shallow, only six paces deep, but that in some channels the depth is so great that anchors do not reach the bottom. Martianus (696) says that the sea there is depressed by deep channels, six paces in depth.[27]

Near the opening of Geometry's discourse, as we have observed, Martianus has the audacity to try to explain Eratosthenes' method of measuring the circumference of the earth. Cleomedes accurately reports the procedures used by Eratosthenes as follows: asuming Syene to be directly beneath the celestial tropic, Eratosthenes measured the length of the Syene-Alexandria arc and, finding the arc of the gnomon's shadow in a hemispherical bowl placed at Alexandria at noon at the summer solstice to be one-fiftieth of a circle, he multiplied the Syene-Alexandria distance by fifty to get a value for the earth's circumference.[28] According to Martianus (597), Eratosthenes measured the Syene-Meroë arc and, determining the length of the gnomon's shadow from the center of the bowl at the equinox, multiplied by

[27] For a partial list of the passages Martianus has garbled from Pliny see Stahl, *Roman Science*, pp. 279-80.

[28] Cleomedes 1. 10.

twenty-four and "got the measure of a double circle."[29] Martianus appears hopelessly confused. Berger suggests that in the words "double circle" (*circuli duplicis*) Martianus had in mind the doubling of the half circle of the bowl.[30] But Remigius glosses *circuli duplicis* as follows: *quia dividitur circulus in duas aequas partes; circulus enim duplex est cuius diameter circulum in aequales partes dividit.*[31] When Remigius says that a diameter divides a circle into two equal parts, he is assuming, as Martianus does elsewhere,[32] that a diameter is half a circle. John Scot Eriugena, commentator on Martianus in another work, devotes several hundred words (three columns of close print in the Migne edition) of his *De divisione naturae* to explaining Eratosthenes' method. John too assumes that a diameter is half a circle.[33] His explanation is an expansion of Martianus' brief statement.

[29] Elsewhere (876) Martianus mistakenly places Meroë at the summer tropic.

[30] E. H. Berger, *Die geographischen Fragmente des Eratosthenes* (Leipzig, 1880), p. 127, also thinks that Martianus meant to use the Syene-equator arc, which, according to Eratosthenes, measured 16,800 stadia and was one-fifteenth of the measurement of Eratosthenes for the earth's circumference. Assunto Mori, "La misurazione eratostenica del grado ed altre notizie geografiche della 'Geometria' di Marciano Capella," *Rivista geografica italiana*, XVIII (1911), 586, thinks that Berger's explanation and correction of Martianus' confused statement is the only possible one. Both scholars are giving Martianus credit for more mathematical skill than he possessed.

[31] Remigius (ed. Lutz, II, 139). "Because a circle is divided into two equal parts; for a circle is 'double' whose diameter divides the circle into equal parts."

[32] Martianus 735 (Dick 370. 13-14), speaking of the number five, says: *Hunc numerum quis neget esse diametrum? Nam decadis perfectio circulusque huius hemisphaerio edissecatur.* [Who would deny that this number represents the diameter? For the decad, representing perfection and the circle, is cut in half by the semicircle of this number.] Remigius *ad loc.* (ed. Lutz, II, 188): *Hunc numerum id est quinarium. Diametrum quasi dimidium denarii vel dimidium circuli.* [I.e., the number five. The diameter is, as it were, half of ten or half of a circle.] For a possible explanation of this mistake, see below, p. 146.

[33] *De divisione naturae* 3. 33 (Migne, *PL*, Vol. CXXII, cols. 716-18): *Duplo enim vincitur ab ipso circulo seu sphaera, eius medietas constituitur. Nam et denarii numeri, veluti cuiusdam circuli, quinarius diametros est.* Later in the passage John gives Eratosthenes' figure for the earth's circumference as 252,000 stadia and says that the diameter is 126,000 stadia, the same figure as the distance to the moon. John was disturbed by the great discrepancy between the figures of two venerable authorities, Eratosthenes (252,000 stadia) and Ptolemy (180,000 stadia), and sought to reconcile them by a difference in standards. See Aubrey Diller, "The

Before beginning his *periegesis* Martianus divides the known world into three continents: Europe, Asia, and Africa. The Strait of Gibraltar separates Europe and Africa, the Don River forms the demarcation between Europe and Asia, and the Nile separates Asia and Africa. These are Pliny's divisions. Martianus concludes by observing (626) that "very many authorities" prefer to regard the Sea of Marmara as the boundary between Europe and Asia. The "very many" are actually Solinus.[34]

Spain gets more than its share of notice from Martianus—as it did from Pliny, the usual explanation for this being that Pliny served as procurator in Spain, and that Varro, his presumed authority, had served as military commander in Spain in the Civil War against Caesar.[35] But Varro and Pliny drew their geography from books and records rather than from personal experience. Pliny also held procuratorships in Gaul and campaigned in Germany but those countries got scant attention from him. Pliny gives undue coverage to Syria, a province which did not figure in his career, if he ever did visit it. Indications are instead that it was the Greek geographers who, just as they bequeathed the framework of the *periegesis* to Roman geographers, determined the amount of discussion to be given to each province of the Empire and to the outer regions of the world. Spain was a land of keen interest to Greek geographers from the time of Hecataeus (*c.* 500 B.C.), while the attention given to Syria probably reflects Posidonius' interest in Pompey's campaigns. Pliny in turn determined the proportionate length of treatment of each region for Latin writers who compiled from him directly or indirectly, such as Solinus, Martianus, Isidore, and Dicuil;[36] and thus, through him, Greek authorities

---

Ancient Measurement of the Earth," *Isis*, XL (1949), 9. Duhem, III, 58-59, comments on John's mistake and sees the traditional figure of 126,000 stadia for the distance to the moon as an important factor.

[34] Cf. Pliny 3. 3 and Solinus 23. 15. Strabo records the divisions given by Pliny as those of Posidonius.

[35] Compare Martianus' account of Spain (627-33) with Pliny's (3. 6-8, 16-18, 21, 29-30) and Solinus' (23. 1-9).

[36] Dicuil, the first Frankish geographer, wrote a survey of the world (A.D. 825) which depends upon Solinus and Pliny and gives no information about Bavaria and Saxony. See *Dicuili Liber de mensura orbis terrae*, ed. G. Parthey, pp. 45-47.

set the form and content of world geography for over a thousand years.

Pliny's place names along the mainland come in helter-skelter profusion. To alleviate the tedium of readers he shows interest and sometimes emotion when writing about great rivers like the Po and Danube or about places with literary or mythological associations; from time to time he retraces his course to remark about offshore islands, including tiny ones, in some detail; and he delights in occasionally inserting an incredible tale about some place. These divertissements had a better chance of survival in Pliny's successors than had his lists of place names. Solinus concentrated on picking out Pliny's *incredibilia* and places with amusing sidelights. Martianus summarizes "memorable features, lauded by the poets," as for Italy (641): the town of Scyllaeum;[37] the Crateis River,[38] mother of Scylla; the whirpool of Charybdis; the rose gardens of Paestum;[39] the cliffs of the Sirens; Campania's glades; the Phlegraean Fields; and Tarracina, dwelling place of Circe. By the seventh century there was not much left of Pliny but the divertissements. Isidore sweeps over entire continents, dropping a few names along the way, but he gives the tiny island of Thanet a whole paragraph in his chapter on the islands of the world because of Solinus' arresting observations that there are no snakes on the island and that earth taken from it to any part of the world kills snakes.[40]

Once past Italy, Martianus picks up speed. Between Italy and Illyricum there are "many peoples, bays, cities, rivers, mountains, and barbarous races" (650)—a glib way of accounting for fifty chapters in Pliny (3. 101-50). Nations are lumped together on the shadowy eastern borders of the northern half of the world—Getae, Dacians, Sarmatians, Wagon-Dwellers, Cave-Dwellers, Alani, Germans. Then the Danube mouth, the Dnieper and Bug Rivers. Beyond are the Geloni, Agathyrsi, Man-Eaters, Arimaspi, and the fabulous Rhipaean Mountains. Across these are the Hyperboreans, living in bliss, and in a trice we are on the

[37] Cf. Homer *Odyssey* 12. 235.

[38] Cf. *ibid.* 124; Ovid *Metamorphoses* 13. 749.

[39] Famed for their twice-blooming roses. Cf. Vergil *Georgics* 4. 119; Propertius 4. 5. 59; Martial *Epigrams* 12. 31. 3.

[40] Cf. Isidore 14. 6. 3; Solinus 22. 8. "Thanet" is derived from the Greek *thanatos* [death].

shores of the Northern Ocean. Less than a hundred words are required to sweep past Germany, Britain, the isles in its vicinity, and Gaul, and to return to Gibraltar (666). Martianus did not have Pliny's stamina and was exhausted by the time he reached Spain. Pliny gave almost as much attention to western Spain and Lusitania on his homeward course (4. 110-20) as he had to eastern Spain on the way out (3. 6-28).

The southern *periegesis* follows immediately, written in the same style, although for this survey Martianus was depending more heavily upon Solinus than upon Pliny. Solinus, purveyor of tall stories, was a matchless figure to follow in recording the wonders of the dark continent and India. He, Pliny, and Martianus were largely responsible for the perpetuation of reports of grotesque monsters down to the time of Stanley and Livingstone. Rudolf Wittkower has written a delightful and well-documented essay in which he states his reasons for believing that an early illustrated Solinus existed and that it is not unlikely that Martianus too was illustrated at an early date.[41] We meet a collection of Pliny's African *monstra* on the way to India (5. 44-46) and another on the way back (6. 190-95). Solinus embellished Pliny's accounts by drawing from Mela as well as from other books of Pliny and occasionally from a source unknown even to Mommsen.[42] Martianus greatly reduces their collections (63-74). He includes the Atlantes, who never dream; the Troglodytes, who dwell in caves, feed on serpents, and hiss rather than speak; the headless Blemmyae, who have mouth and eyes in their chests; and the Strap-Feet, who crawl because their feet are maimed.

Martianus takes greater care in excerpting from Pliny in the section on Numidia (669-70), probably because it was his homeland. This account contains his own remark about Carthage, which, as we saw above, is used in setting a *terminus ante quem* for his book. He repeats the current notion, which Pliny got from King Joba's book on Libya,

[41] "Marvels of the East," *JWCI*, V (1942), 159-97. See esp. "The Pictorial Tradition," pp. 171 ff. Wittkower traces thirteenth-century Italian miniatures illustrating Solinus to a sixth- or seventh-century archetype. See also Kurt Weitzmann, *Ancient Book Illumination* (Cambridge, Mass., 1959), pp. 18, 142, n. 57.

[42] Solinus 30. 1 - 31. 6 (ed. Mommsen, pp. 130-37). Mela's accounts of African monsters are sometimes fuller than Pliny's. For a comparison of texts see D. Detlefsen, *Die Geographie Afrikas bei Plinius und Mela und ihre Quellen* (Berlin, 1908), pp. 35-40.

that the Nile originates in a lake in Mauretania.[43] Palmyra, between Syria and Parthia, is mentioned (681). The city had been destroyed by the Romans A.D. 273; but this has no bearing upon Solinus' or Martianus' dates, because the item comes from Pliny. It is pointless to detail all of Martianus' mistakes and garbling in his reductions of Pliny and Solinus. The parallel passages are noted in the apparatus of Dick's edition.[44] At one point the relentlessly sharp Hugo Grotius, by collating Solinus' reading, corrected Martianus' mistake of putting Lycaonia in the nominative case (686). But, as Kopp wryly asks in a note on this passage in his edition, is it our intention to correct Martianus or the manuscripts?[45]

Soon we come to India (694-96), a land that calls for a full treatment, and Ceylon (696-98), a sort of Ultima Thule of the East. India has five thousand cities and is thought to be one third of the world. Kings rule here, and there is an abundance of armies and elephants. Indians beautify themselves by dyeing their hair, some with bluish, others with yellowish, dyes; they adorn themselves with jewels and consider it a distinction to ride on elephants. Ceylon is gigantic in size—7,000 stadia in length and 5,000 in breadth—and also in its contents: elephants and pearls are larger there than in India, and the men are larger than humans anywhere. A man who dies at the age of one hundred dies prematurely. According to Martianus, the Ceylonese have no oral dealings with outside peoples but employ the method of dumb barter at a river bank for exchanging wares. What Pliny actually said (6. 88) was that this was a Chinese custom, reported by Ceylonese ambassadors. Is it any wonder that Martianus was an esteemed geographer in the Middle Ages?

As might be expected, the homeward swing around Africa is even swifter than the circumnavigation of northern shores. Skirting the southern coast of Africa was no problem to Pliny; it was merely a matter of opinion: the Island of Cerne lies off Ethiopia, according to

[43] On the notion of the Nile's rising in an oasis south of Mt. Atlas, see Thomson, pp. 70-71, 267-69.

[44] Dick omits many Solinus parallels, but most of these can be recovered by consulting the index in Mommsen's edition of Solinus.

[45] Leonardi suggests some emendations, drawn from Rather's glosses, for the text of Book VI; see "Raterio e Marziano Capella," pp. 88-89.

Ephorus; but Polybius states that it lies on the edge of Mauretania, opposite Mt. Atlas (6. 198-99). So it is with Martianus (702). In one sentence we are in the parched interior of Ethiopia, in the next on the Gorgades Islands and the Isles of the Blessed, across from Mauretania. We can accept Geometry's concluding statement (703) that she could not tarry on such a journey, but we cannot accept her weak excuse that the places she has skimmed are insignificant (*ignobilia quaeque praetervolans*). The course of history might have been radically altered if Geometry had deigned to give us an account of her African travels.

Even the place names of Martianus, tedious as they are when they occur in clusters, had an influence upon medieval cartography.[46] Four of the ten known manuscripts of the *Liber floridus* of Lambert of St. Omer (*fl.* 1120) contain a *mappa mundi*. Richard Uhden made a careful examination of the map in the Wolfenbüttel manuscript of the *Liber* and ascertained[47] that it was not related to Lambert's work, that instead one of the legends on the map ascribes it to Martianus, and that various other legends, including the one for the antipodes, bear unmistakable similarities to the text of Martianus. Uhden found that 57 per cent of the names on the map occur in Martianus. Another careful study of maps and accompanying glosses was made by Leonardi, who believes that the two cartographic drawings found in Codex San Marco 190 are the oldest that are traceable in a Martianus codex.[48]

At the beginning and the end of his *periegesis* Martianus gives the over-all dimensions of *terra cognita* (609-16, 703). From Pliny he got the correct figure for Eratosthenes' measurement of the earth's circumference—31,500 miles[49]—and he also gives the correct figure for Ptole-

[46] Including the celebrated Hereford Map. See Beazley, I, 341.

[47] "Die Weltkarte des Martianus Capella, *Mnemosyne*, III (1936), 97-124. Konrad Miller, *Die ältesten Weltkarten*, III, 50, recognized that this map was based on an earlier prototype than was Lambert's work. Miller's transcribed copy of the Wolfenbüttel *mappa mundi* is inaccurate. For a clear facsimile of the map see Youssouf Kamal, *Monumenta cartographica Africae et Aegypti*, III, 777.

[48] "Illustrazioni e glosse in un codice di Marziano Capella," *Bullettino dell'Archivio paleografico italiano*, new ser., Vols. II-III, pt. 2 (1956-1957), 39-60.

[49] There are 125 Roman paces in a stadium (Pliny 2. 85), 1000 paces in a Roman mile (*millia passuum*); 252,000 stadia = 31,500 Roman miles (Pliny 2. 247). D. R. Dicks (ed.), *Hipparchus' Geographical Fragments* (London, 1960), pp. 42-46, 150-52, points out the difficulties, if not the impossibility, of obtaining an accurate value for the stadium.

my's estimate in both stadia (180,000) and Roman miles (22,500). We may be sure that Martianus got Ptolemy's figure from some Roman intermediary and not from Ptolemy's *Geography* (1. 11. 2), as has been commonly supposed.[50]

The longitudinal distance from the eastern extremity of India to Cádiz is 8,578 miles (611-12), according to Artemidorus. Martianus got the figure and attribution from Pliny (2. 242). The figures of Pliny, as given in the manuscripts, are in Roman numerals, which are frequently incorrectly copied by scribes. Martianus' quantities are written out in Latin, and, since he was using older manuscripts of Pliny and Solinus than those we possess, it is obvious that Martianus' text must be included in any collation of manuscripts for Pliny's geographical books.[51]

Martianus then states (611) that the distance as recorded by Isidore[52] is 9,818 miles, and, in attempting to reconcile the discrepant estimates, he points out that Artemidorus meant to add to his figure the distance from Cádiz to Cape Finisterre, 991 miles. This too was obtained from

[50] Ammianus Marcellinus is the only Latin writer that I know of before the period of Arabic influence who assuredly read Ptolemy's *Geography*, and he was a Syrian Greek who rather surprisingly chose to write in Latin. Martianus nowhere indicates an inclination to peruse a technical work like Ptolemy's. Leonardi's remarks about discrepancies between Ptolemy and Martianus ("Nota introduttiva," pp. 274-75) are wholly irrelevant. Martianus and Solinus were not familiar with Ptolemy traditions. Ptolemy's figure originated with Posidonius and was probably available to Pliny in Varro's writings. See I. E. Drabkin, "Posidonius and the Circumference of the Earth," *Isis*, XXXIV (1943), 509-12.

[51] Martianus (611) writes out the quantity as eight thousand five hundred and seventy-seven. Rackham, editor of the Loeb edition of Pliny, was unaware of the Martianus correspondences. Rackham reads 8,568, and his figures for the intervening distances do not give a correct total. Jean Beaujeu, in the Budé edition of Pliny, II, 109, reads 8,578, and his figures for the intervening distances do give a correct total. Rackham's edition is marred by careless mistakes, mistranslations, and misreadings of the manuscripts. It is hoped that the long-awaited Budé edition of Pliny, Books III-VI, will soon appear. It ought to clear up many of the faulty readings of Pliny's figures. On the problems of manuscript transmissions of Pliny's numerals see Beaujou, pp. 244-46.

[52] Isidore of Charax, one of whose works on overland distances, *The Parthian Stations* (*c.* A.D. 25), has survived. The Greek text was edited and translated, with commentary, by Wilfred H. Schloff (Philadelphia, 1914).

Pliny. Geometry's own estimate of the distance by land and sea, corroborated by Artemidorus, is 8,685 miles.[53]

Martianus gives the latitudinal distance, from the shores of the Ethiopian Ocean to the mouth of the Don River, as 5,462 miles.[54] That estimate is reduced by 678 miles, if made over the seas.[55] According to Artemidorus the mouth of the Don marked the upper limit of geographical knowledge, but Isidore calculated another 1,250 miles—the distance to Ultima Thule—an unreliable conjecture, in Martianus' opinion (616). This figure of 1,250 miles, its attribution to Isidore, and the skeptical attitude all come from Pliny, but Pliny had good reasons for being skeptical. Martianus prefers an absurd explanation, perhaps his own, that the earth was earlier shown to be spherical and it is impossible for a sphere to have unequal sides.

At this point Geometry is a pitiful sight, winded from the exertions of her long recitation. Venus has been frowning through it all; this is not her idea of a wedding celebration. Desire, one of Venus' attendants, lashes out at the maiden unmercifully, calling her senseless and boorish, her limbs so roughened by traipsing over mountains, straits, and highways, and her appearance so shaggy that she could be taken for a man. Nevertheless she is directed to get on with her discourse on geometry, and to be brief about it (704-5).

Geometry's ten-page digest of Euclidean elements is, for its time and place, a remarkable treatment of the subject. It is also a document of some significance among the scant vestiges of Euclid in the Latin world of the early Middle Ages.[56]. It resembles a Greek systematic

[53] Martianus (613) writes out the total distance in Latin. Pliny (2. 244-45) gives distances between intervening stations, an impressive list, especially since some distances are given in half miles, but the figures adopted by Rackham and Beaujeu do not add up to the totals, and Pliny's totals in both editions are different from Martianus'. On Pliny's errors see Miller, VI, 135-40.

[54] Again Martianus writes out the total distance and repeats Pliny's list of intervening stations, but he does not give the distances between stations. The figures adopted by Beaujeu and Rackham differ, and neither adds up to Pliny's (2. 245) total, which is the same as Martianus'.

[55] Martianus writes out the quantity. Beaujeu reads 79 miles; Rackham omits a figure.

[56] Martianus is given some slight weight in questions of the authenticity of Euclid's text. See *The Thirteen Books of Euclid's Elements*, ed. Sir Thomas Heath, I, 62, 187

handbook,[57] exhibiting a reasonable sense of order, and containing definitions and divisions of the subject. We may safely assume, from our knowledge of the habits of Martianus, that he did not make his own compilation of extracts from a larger Euclidean work but appropriated some already prepared digest, examples of which are discussed by Tannery,[58] Heath, [59] Goldat,[60] and Ullman,[61] and texts of which are found in Tannery,[62] Bubnov,[63] the Friedlein edition of Boethius' mathematical writings,[64] the *Corpus agrimensorum*,[65] Goldat's edition of a twelfth-century version of the *Elements* (Bibliothèque Nationale, Fonds latin 10,257)[66] and Geymonat's edition of the Verona fragments.[67] Martianus includes many of the traditional rudiments found in medieval digests and extracts of the *Elements*: the definitions, five postulates, and first three axioms of Book I; classifications of angles and of plane and solid figures; and a few definitions from Books V,

[57] See above, "The Work," n. 48.

[58] Most of Paul Tannery's writings on geometry in the Latin West have been collected in his *Sciences exactes au moyen âge*, ed. J. L. Heiberg, and are still worth perusal. See "La Géométrie au XIe siècle," pp. 79-102, esp. pp. 95-100; and "Notes sur la pseudo-géométrie de Boèce," pp. 211-28.

[59] See his edition of Euclid's *Elements*, Vol. I, Introduction and chap. viii.

[60] *The Early Medieval Traditions of Euclid's* Elements discusses the geometry in the *Corpus agrimensorum* (pp. 32-39) and the authorship of the "Boethian" versions of Euclid (pp. 54-59). Goldat also analyzes the contents of the *Ars geometriae et arithmeticae Boetii* in five books.

[61] Ullman, p. 265, discusses Balbus' *Expositio et ratio omnium formarum* (2d cent. A.D.); the Codex Arcerianus (Wolfenbüttel), which Tannery dated in the sixth to seventh centuries but is now dated in the fifth or sixth (pp. 267-72); the two geometries ascribed to Boethius; and the then forthcoming Geymonat edition (now published; see n. 14 above) of the Verona fragments (p. 272). Ullman also provides an up-to-date bibliography.

[62] "Un Nouveau Texte de traités d'arpentage et de géométrie d'Epaphroditus et de Vitruvius Rufus," in *Sciences exactes au moyen âge*, pp. 29-78.

[63] In his edition of Gerbert's *Opera mathematica*, pp. 494-508.

[64] See above, n. 12.

[65] F. Blume, K. Lachmann, A. Rudorff, eds. *Gromatici Veteres: Die Schriften der römischen Feldmesser* (2 vols., Berlin, 1848-1852); Carl Thulin, ed., *Corpus agrimensorum romanorum* (edition not completed), Vol. I (Leipzig, 1913).

[66] Goldat, pp. 88 ff.

[67] See above, n. 13. L. W. Jones has included another medieval excerpt, found in a manuscript of Cassiodorus' *Institutiones*, in his translation of that work: *An Introduction to Divine and Human Readings*, pp. 216-19.

X, and XI. Also included are some matters that are not found in Euclid; and, at times, as we shall see, Martianus follows the definitions of Heron of Alexandria more closely than those of Euclid. The traditions of Euclid and Heron dominated medieval geometry. It is likely that the immediate antecedent of Martianus' Book VI presented a conflation of both traditions, just as the antecedent of Book VII probably presented a conflation of Euclidean and Nicomachean arithmetic.

A suitable approach to a treatise of this sort is to compare it with the *Ars geometriae*, a compendium supposedly translated from Euclid and falsely attributed to Boethius, which, as Ullman has recently shown, turns up with great frequency in copies or extracts in medieval library collections of codices on geometry and surveying.[68] But since there are also indications of a Heronic tradition in Martianus' digest, it would be well to keep a copy of Heron's *Definitiones*[69] at hand for comparisons.

Some opening remarks about the derivation of lines from incorporeal points and of numbers from the indivisible monad (706-7) bear a striking resemblance to Macrobius' dicussion of corporealities and incorporealities[70] and serve to show the close relationship between Geometry and her sister Arithmetic. Then follows the division of figures into plane (*epipedon*) and solid (*stereon*), the former originating in a point (*semeion*), the latter in a surface (*epiphaneia*).[71] Next

[68] Ullman, pp. 270-71, prefers to regard the *Ars geometriae* as condensed, rather than spurious, Boethius. The text is found in the Friedlein edition of Boethius' mathematical works, pp. 372-428. Friedlein regarded the work as spurious, which is the view usually adopted by historians of mathematics. See Heath's edition of the *Elements*, I, 92; Heath, *History of Greek Mathematics*, I, 359-60. The work contains proofs of Book I, Props. 1-3, evidence, according to Heath, that the pseudo-Boethius had the use of a Latin translation of Euclid. (Heath of course meant to say a translation of at least portions of Euclid.) Marshall Clagett, *Greek Science in Antiquity*, p. 151, calls the translation a "lamentable rendering." See the discussions of this treatise in Tannery, *Sciences exactes*, pp. 97-99, 211-18, 246-50.

[69] Vol. IV of Wilhelm Schmidt's critical edition of *Heronis Opera*, ed. J L. Heiberg.

[70] Macrobius *Commentary* 1. 5. 5-7; and for other parallels see my translation, pp. 96-97.

[71] Cf. *Ars geometriae* (ed. Friedlein 403. 14-24); Gellius 1. 20. 1-2; Cassiodorus 2. 6. 2; Isidore 3. 11. 1-2; 3. 12. 1. Martianus' free use of Greek terminology

come the definitions from Book I of Euclid. Martianus omits Definitions 4, 7, and 21; the *Ars geometriae* gives a fairly literal translation of all twenty-three definitions but inverts the order of Euclid's 13 and 14 (Martianus does not invert them). Martianus mistranslates Definition 1: "A point is that whose part is nothing," instead of "A point is that which has no part."[72] After giving Euclid's definition (2) of a line, Martianus interposes definitions of the four kinds of lines—straight, circular, spiral-shaped, and curved—as does Heron (*Definitiones* 3-7). The *Ars geometriae* defers definitions of three kinds of lines—straight, circular, and curved—to a later part of the book (Friedlein ed. 394. 2-14). Euclid omits a classification of species of lines.[73] Martianus amplifies Euclid's definition of a surface, adding that it lacks depth, as does color on a body, and that the definition applies to both plane and curved surfaces. Euclid defines only a plane surface. Following his translation of Definition 18 of Euclid, Martianus interposes a classification and definitions of three kinds of plane figures: those contained by straight lines (*euthygrammos*), those contained by

---

throughout the quadrivium books does not necessarily support the conjecture that he was translating from Greek sources. Greek technical terms, sometimes transliterated, but in Martianus usually kept in Greek characters, were part of the Latin tradition of Greek mathematical writings. In general the retention of Greek characters does indicate a higher level of writing than the transliteration of the characters.

[72] Heath (*Euclid*, I, 155), thinks that Martianus' mistranslation may be unique. However, Goldat, p. 35, n. 108, points out that Cassiodorus' *Expositio in psalterium* (Migne, *PL*, Vol. LXX, col. 684) has the same faulty definition. Heath (*Euclid*, I, 91), thinks that the occurrence of several Greek terms in Martianus' epitome indicates that he was using a Greek Euclidean source and that he may have been to blame for the mistranslation. Martianus probably could read Greek well enough to have derived his brief digest from some Euclidean primer, but a much more likely assumption is that he was following a Latin Euclidean tradition which preserved the Greek terms and which may have originated in Book IV of Varro's *Nine Books of the Disciplines*. Varro is known from the Gellius quotations to have used Greek terminology, and Greek Euclidean primers would have been more readily available and of greater interest to students in Varro's day than in Martianus'.

[73] Heath (*Euclid*, I, 159), says that he omitted the classification because it was not necessary for his purpose. On classifications of lines by Heron and others see pp. 159-65.

curved lines (*kampulogrammos*), and composite plane figures, contained by both straight and curved lines (mikton).[74] Euclid does not classify surfaces in general. He does, of course, distinguish between regular and nonregular polygons and polyhedra.

Martianus paraphrases (712) Euclid's definitions of rectilinear figures (Defs. 19-22) and treats them as a classification of the *euthygrammos* species. He then proceeds (713) to the second species, figures contained by curved lines, and divides them into circular and elliptical figures; and to the third or "mixed" type, represented here by the semicircle. He had previously defined (711) the semicircle as "a figure that is contained by the diameter and the circumference which that same diameter cuts off in the middle."[75] His own addition of the adjective *media* [half] to modify *peripheria* may be related to his and subsequent Carolingian misapprehensions that a diameter is half a circle.[76]

Martianus next deals with the two kinds of geometric propositions: problems and theorems. He first defines (715) the Greek terms applied to the steps involved in constructing figures: *tmēmatikos*, "cutting lines to a prescribed length"; *sustatikos*, "joining given lines"; *anagraphos*, "describing a figure upon a line"; *engraphos*, "enclosing a prescribed triangle or some other figure within a given circle"; *perigraphos*, "circumscribing a quadrate or some other figure about a circle"; *parembolikos*, "inscribing a given triangle within a given tetragon"; and *proseuretikos*, "finding a line that is the mean proportional between two given lines." This classification seems to be unique in the extant writings on geometry.

Moving on to terms applying to theorems, Martianus presents (716) these five formal divisions of a proposition: *protasis* [enunciation], *diorismos* [definition], *kataskeuē* [construction], *apodeixis* [proof], and *sumperasma* [conclusion]. He omits *ekthesis* [setting out], included by

[74] Cf. Heron's classification (Def. 74) into incomposite and composite or, again, into simple and mixed. See Heath's edition of the *Elements*, I, 170.

[75] *Hemicyclum est figura, quae diametro et peripheria media, quam eadem diametros distinguit, continetur.* Cf. *Ars geometriae* (ed. Friedlein 375. 11-13): *Semicirculus vero est figura plana, quae sub diametro et ea, quam diametrus apprehendit, circumferentia continetur.*

[76] See above, nn. 31 and 32.

Proclus in his *Commentary on Euclid I*;[77] and *dissolutio* [reduction to absurdity], included in the Adelard III version of Euclid.[78]

Martianus feels that a further discussion of angles is needed (717). Once again there are three kinds: right angles are always the same and equal; acute angles and obtuse angles are always "variable" (*mobilis*)—any angle broader than a right angle is obtuse; a narrower angle is acute. The compiler of the *Ars geometriae* felt a similar need to elaborate later on the three kinds of angles.[79]

Then follow the definitions of four terms that are used to describe proportional relationships, garbled or reduced in the Latin transmissions of Euclid. *Isotēs* [equality]: two lines of equal length are compared with a third that is of double or equal length.[80] *Homologos* [corresponding]: magnitudes compared are "in agreement" (*collata consentiunt*).[81] *Analogos* [proportional]: a line half as long as another is twice as long as a third. *Alogos* [disproportional]: lines neither are equal nor bear any other rational relationship to each other.[82]

Next Martianus defines and briefly comments on (718) commensurable and incommensurable lines and lines commensurable "in power" or "in square," as Heath prefers to translate the Euclidean term.[83] Then follows a list (720) of the thirteen kinds of irrational straight lines, in the order in which they are found in Euclid.[84]

Martianus' discussion of solid figures is very brief (721-22). He gives Euclid's definitions of a solid and its extremities,[85] and points to the generation of a pyramid from a triangle, a cone or a cylinder from a

[77] Ed. G. Friedlein (Leipzig, 1873), pp. 203-4.

[78] See M. Clagett, "King Alfred and the *Elements* of Euclid," *Isis*, XLV (1954), 272.

[79] Friedlein ed., 393. 11 - 394. 1. M. Fuhrmann, *Das systematische Lehrbuch*, points out *passim* that it was standard practice of *Lehrbuch* authors first to classify and define and later to elaborate.

[80] I.e., given $A=B$, either $A+B=C$ or $A=B=C$ is considered a relationship of equality.

[81] A reduction of Euclid (*Elements* 5. Def. 11), the interpretation of which has given commentators difficulty. See Heath's edition of the *Elements*, II, 131.

[82] Cf. Euclid *Elements* 5. Defs. 17, 11, 9.

[83] Cf. Euclid 10. Defs. 1, 2. And see Heath's edition, III, 11.

[84] Euclid 10. Prop. 111.

[85] Cf. Euclid 11. Defs. 1, 2.

circle, a cube from a quadrate, and a sphere from a circle. To these are added the "noble figures" (*nobilia schemata*), fashioned from the others: the octahedron, dodecahedron, and icosahedron.[86]

Martianus' digest of Euclidean elements concludes with translations of the five postulates and of the first three of Euclid's five axioms for Book I. Heron gives only three axioms. The *Ars geometriae* gives the first four and adds to the group Definitions 1 and 2 of Book II of Euclid.[87]

Her lessons in Euclidean principles concluded and accepted, Geometry draws a straight line on her abacus and asks how one goes about constructing an equilateral triangle upon a given straight line (724). The philosophers[88] in the audience immediately recognize that she is preparing to work out the construction for the first proposition of Euclid and they break into acclaim of Euclid. Geometry, pleased with this approbation of her disciple, snatches from his hand his precious books—which undoubtedly contain the proofs of all the propositions in the thirteen books of the *Elements*[89]— and gives them to Jupiter as a textbook for the further instruction of the heavenly company.

[86] Primary bodies, according to Plato *Timaeus* 55 b-c. And see the note in the Cornford edition.

[87] On the question of the authenticity of Euclid's axioms see Heath's edition, I, 62, 221-22.

[88] In late antiquity the term "philosopher" was applied to men in all branches of learning, including such technical fields as mining engineering. See Curtius, pp. 209-11.

[89] Whether a complete translation of the *Elements* was available in the West before the twelfth century—the basic question for all investigators of Latin geometry—has not yet been answered decisively. As Marshall Clagett ("The Medieval Latin Translations from the Arabic of the *Elements* of Euclid," *Isis*, XLIV [1953], 16-17) points out, investigators have been unable to prove the existence of a complete *Elements*. According to Goldat (p. 37), the available evidence does not indicate that there was a complete translation of the *Elements* into Latin before the Arabic versions became known to the West in the twelfth century. At any rate, it is certainly clear that for Martianus Euclid was a legendary figure, one of "the ancients" (*veteres, antiqui*), whose reputation owed more to tradition than to genuine comprehension or appreciation.

On Arithmetic

ONCE Geometry has ended her discourse, we anticipate the entrance of Arithmetic, a sister of Geometry—in fact her closest sister. A little earlier (706-7) Geometry expatiated on the binding ties between the two, observing that all assertions about matters which progress to infinity are expressed either in numbers or in lines. Numbers are apprehended by the intellect, lines by the sight. Numbers belong to Arithmetic; linear figures, the province of Geometry, are demonstrable on her abacus. Lines are begotten of incorporealities and are fashioned into manifold perceptible shapes "that are even elevated to the heavens".[1] The beginnings of lines are incorporeal and are shaped by both sisters. The indivisible monad is the begetter of numbers, itself not a number, and is not apprehensible. The monad also represents an indivisible Euclidean point, which is likewise not apprehensible. Numbers are incorporeal, unless they are applied to objects. Thus when Arithmetic is introduced, the attendants are requested not to remove Geometry's abacus.

As she enters the celestial hall, Arithmetic is even more striking in appearance than was Geometry with her dazzling peplos and celestial globe. Arithmetic too wears a robe, hers concealing an "intricate undergarment that holds clues to the operations of universal nature."[2] Arithmetic's stately bearing reflects her pristine origin, antedating the

[1] Geometry's obscure remark is intended to show that astronomy is also closely related to geometry and arithmetic. When Geometry was first introduced (580-81), she was wearing a peplos covered with geometric figures that "served the purposes of her sister Astronomy as well." On the Platonic and Euclidean character of Greek astronomical theory see the excellent discussion of E. J. Dijksterhuis, *The Mechanization of the World Picture*, pp. 54-68.

[2] *Huius autem multiplicem pluriformemque vestem quoddam velamen, quo totius naturae opera tegebantur, abdiderat* (729). The robe may symbolize pure numbers, and the undergarment their application to the physical world. At least this is the explanation of Remigius *ad loc.* (ed. Lutz, II, 178).

birth of the Thunder God himself.[3] Her head is an awesome sight. A scarcely perceptible whitish ray emanates from her brow; then another ray, the projection of a line, as it were, coming from the first. A third ray and a fourth spring out, and so on, up to a ninth and a tenth ray—all radiating from her brow in double and triple combinations. These proliferate in countless numbers and in a moment are miraculously retracted into the one.[4] The acumen of Remigius is not required to refer this symbolic description to the sacred Pythagorean decad.[5] The first ray represents the monad or the point, the source of all numbers or all geometric figures. The decad encompasses all numbers. It and all numbers confined within it, "in double and triple combinations," were worshiped as sacred by Pythagoreans.[6]

Arithmetic's fingers vibrate with a speed that blurs the vision. She is calculating,[7] and the sum she produces is 717. Philosophy explains that Arithmetic, in this computation, is greeting Jupiter by his very own name (729).[8] The countless rays that spring from her brow frighten some of the earth-born deities standing by, and, fancying that Arithmetic is sprouting heads like the Hydra, they look to Hercules for help. At this moment Pythagoras, who is the patron and authority of this book, as Euclid was of the last, accompanies Arithmetic to the abacus and holds a torch above her head as she proceeds to her discourse. The book on arithmetic that follows originated in two works

[3] 728. Arithmetic is also prior to numbers. Jupiter is later asked by Arithmetic to acknowledge her as his source (731-32). Jupiter is identified with the monad, the source of all numbers (731). See above, p. 144.

[4] *Nam primo a fronte uno sed vix intelligibili radio candicabat, ex quo item alter erumpens quadam ex primo linea defluebat. Dehinc tertius et quartus tuncque etiam nonus decuriatusque primus honorum reverendumque verticem duplis triplisque varietatibus circulabant. Sed innumerabili radios multitudine prorumpens in unum denuo tenuatos miris quibusdam defectibus contrahebat* (728).

[5] Remigius *ad loc.* (ed. Lutz, II, 177).

[6] *Nicomachus* (tr. D'Ooge), pp. 88-108, on the decad and the associations and attributes of the numbers within it.

[7] On what is known about finger-reckoning in the ancient world see n. 49 below.

[8] Perhaps the best explanation of the identification of this number with Jupiter's name is offered by Remigius *ad loc* (ed. Lutz, II, 179): Jupiter to the Greeks was known as Η Α Ρ Χ Η [The Beginning]. The numerical values of the Greek letters are: Η=8; Α=1; Ρ=100; Χ=600; Η=8; total=717.

that Martianus did not consult directly: the *Introduction to Arithmetic* by Nicomachus of Gerasa (*c.* A.D. 100);[9] and Books VII-IX of Euclid's *Elements*, the books dealing with arithmetic (see above, pp. 48-49).

The Greeks recognized two divisions of the subject of arithmetic, both being included in Arithmetic's presentation: arithmology and arithmetic proper. Arithmology, a study of the mystical properties of the numbers in the decad, dealt with the attributes, epithets, and magical powers of these numbers, which it identified with the gods and with a variety of animate and inanimate objects. Arithmetic proper was a rigid mathematical discipline, relating to the properties and theory of numbers, and involving proofs.[10] Naturally treatises on the magical properties had wider circulation than those on the mathematical properties of numbers. Many of the writers on number dealt only with arithmology.[11] It is to the credit of Martianus that he gives

[9] Anyone interested in the content and development of Greek arithmetic should consult the excellent introductory chapters and commentary by F. E. Robbins in the volume containing the D'Ooge translation of Nicomachus. It is clear that Martianus was using intermediate sources, perhaps Apuleius' lost translation of the *Introduction to Arithmetic* for his Nicomachean arithmetic, more likely a late digest of this or some other Latin translation. For his Euclidean arithmetic he used some Latin digest, as we shall see.

[10] Robbins (*Nicomachus*, tr. D'Ooge, p. 28) distinguishes between authorities like Euclid, who offer proofs, and descriptive writers like Nicomachus, who do not. See Heath, *History*, I, 98, on the differences between these methods.

[11] Robbins (*Nicomachus*, tr. D'Ooge, pp. 90-91) lists thirteen accounts of arithmology, including three books devoted wholly to the subject, that have survived in more or less complete form. Nicomachus' *Introduction to Arithmetic* deals only with arithmetic proper. He composed a separate work on arithmology, entitled *Theologumena arithmeticae*, the contents of which are known to us from a digest made by the Byzantine scholar Photius (*fl.* A.D. 870) and from an anonymous treatise with the same title, thought to have been derived largely from Nicomachus' treatise and commonly attributed to Iamblichus. Robbins (*Nicomachus*, tr. D'Ooge, p. 82) considers the attribution a likely one; the latest editor of the anonymous *Theologumena arithmeticae* (Leipzig, 1922), V. de Falco, designates the author as pseudo-Iamblichus. See *Nicomachus*, tr. D'Ooge, pp. 82-87, for an account of Nicomachus' lost *Theologumena*. Heath, *History*, I, 97, regards the anonymous treatise as a conflation, only partially derived from Nicomachus' lost treatise.

much greater attention to arithmetic (743-801) than to arithmology (731-42).

We find it hard to understand why scientists in the ancient world gave serious thought to arithmology. The same can of course be said about astrology. Treatises on these subjects strike us as curiosities, and the recital in ten more pages of the epithets, attributes, and powers of ten numbers becomes for us tedious reading. A digest of Martianus' whole treatment of arithmology will not serve our purposes as well as a presentation in detail of Arithmetic's praise of one number, seven, as a specimen.[12] This number receives the fullest account in other extant treatises and offers the best opportunity to compare Martianus' treatment with that of other writers. Gellius happens to preserve Varro's chapter on the number from a lost work, and Macrobius devotes to seven the longest chapter in his *Commentary*.[13]

Arithmetic asks why the heptad is venerated and straightway provides answers in abundance. It is the "virgin" number and is called Minerva because it shapes the works of nature without any procreative contact. It is the only number that begets no number and is begotten by no number within the decad.[14] Being the sum of male and female numbers three and four, it is named for the mannish goddess Athena.

[12] There is perhaps no subject in the literature of ancient compilers in which verbatim copying and close borrowing are more in evidence than in the extant arithmologies. For a digest of the contents of a typical arithmology see *Nicomachus*, tr. D'Ooge, pp. 104-7.

[13] W. H. Roscher, *Die hippokratische Schrift von der Siebenzahl*, makes an exhaustive study of this number in arithmological literature. Both Roscher (*ibid.* and *Hebdomadenlehren der griechischen Philosophy und Ärzte* [Leipzig, 1906]) and Robbins ("The Traditions of Greek Arithmology," *CP*, XVI [1921], 97-123, and "Posidonius and the Sources of Pythagorean Arithmology," *CP*, XV [1920], 309-22) offer parallel passages which reveal close borrowing and one instance of six pages of verbatim copying. Robbins (*Nicomachus*, tr. D'Ooge, p. 106) gives reasons for the prominence of the number seven in these treatises. On the archaic, seemingly pre-Pythagorean character of the doctrine of the hebdomads see C. J. de Vogel, *Pythagoras and Early Pythagoreanism*, pp. 168-74.

[14] I.e., it is the only prime number that is not a factor of another number within the decad. According to Greek mythology Athena was the virgin goddess and sprang in full panoply from the brow of Zeus. The epithets of the number seven and the reasons for them are given by Macrobius *Commentary* 1. 6. 11; for parallels in the arithmological literature see the *Commentary*, tr. Stahl, pp. 101-2.

Seven refers to celestial phenomena. There are seven phases of the moon: crescent, half (first quarter), waxing gibbous, full, waning gibbous, half (last quarter), and crescent.[15] A lunar cycle is 28 days long,[16] and 28 is the sum of 1, 2, 3, 4, 5, 6, 7. There are also seven celestial circles,[17] seven planets,[18] seven days in a week,[19] and seven transmutations of the elements: formless matter into fire, fire into air, air into water, water into earth, earth into water, water into air, and air into fire.[20] Fire is not transmuted into imperceptible matter. (738)

Lastly, seven controls man and his development. Seven-month parturitions are the first to produce living offspring.[21] Man has seven openings in the head, which provide him with his senses.[22] His first teeth appear in infancy in the seventh month,[23] his second in the seventh year.[24] The second hebdomad of years brings puberty and the power to produce offspring;[25] the third covers his cheeks with a beard;[26] the fourth marks the end of his increase in stature;[27] in the fifth he reaches the peak of his physical powers.[28]

[15] Cf. Macrobius 1. 6. 54-56.

[16] Cf. Macrobius 1. 6. 52; Gellius 3. 10. 6. This approximate figure suited the arithmological doctrines. Later (865) he somewhat more accurately records a sidereal month as $27^2/_3$ days and a synodic month as $29^1/_2$ days.

[17] Seven is not commonly associated with the celestial circles, which are variously counted as eight, nine, ten, or eleven. The complete list numbers eleven: five parallels of latitude, two colures, the ecliptic, the zodiac, the horizon, and the Milky Way. Martianus here must be counting the celestial parallels as seven, as Varro did, according to Gellius 3. 10. 3.

[18] Cf. Macrobius 1. 6. 47; Gellius 3. 10. 2.

[19] This association too is uncommon in classical arithmologies. The seven-day week did not become official in the Roman Empire until Constantine adopted the Christian week for the Roman civil calendar.

[20] Cf. Macrobius 1. 6. 32-33, 36.

[21] Cf. Macrobius 1. 6. 14; Gellius 3. 10. 8; Cicero *De natura deorum* 2. 69, and note in Pease edition.

[22] Cf. Macrobius 1. 6. 81.

[23] Cf. Macrobius 1. 6. 69; Gellius 3. 10. 12.

[24] Cf. Macrobius 1. 6. 70; Gellius 3. 10. 12.

[25] Cf. Macrobius 1. 6. 71.

[26] How contrived this scheme of hebdomads is is indicated by the fact that some writers—e.g., pseudo-Iamblichus—assign the beard to the second.

[27] Macrobius (1. 6. 72) adds increase in breadth.

[28] Cf. Macrobius 1. 6. 72.

Those familiar with arithmological treatises will wonder why Martianus here omits the sixth and seventh hebdomads of years, regular features in other accounts. Leonardi appears to provide the answer: Isidore of Seville (*De numeris* 188c-d), who is copying from this passage of Martianus, includes a sixth hebdomad, marking deterioration, and a seventh, marking the beginning of old age. Leonardi believes that there was a lacuna here in the archetype of the extant manuscripts of Martianus but not in the manuscript used by Isidore.[29]

Man's vital organs also number seven: tongue, heart, lung, spleen, liver, and two kidneys.[30] And the members of his body are seven: head (including neck), chest, belly, two hands, and two feet.[31] For full measure Arithmetic adds that there are seven stars at the celestial pole.[32] (739)

By Martianus' time arithmetic had taken precedence over geometry in the curriculum of the Latin schools.[33] From the beginning the Romans were interested only in the practical applications of mathematics and were never attracted to Greek mathematical theory. They looked to geometry for its adaptability to surveying, and to arithmetic as an aid to computation. While geometry was becoming a technical specialty, the attractions of arithmetic were steadily being enhanced by several developments. Under the Roman emperors the Latin world shared in the revival of Pythagoreanism, with its number symbolism; and the last pagan school of philosophy, Neoplatonism, was intimately linked with Neopythagoreanism.[34] The Christian Fathers, after the Church's victory over paganism, placed arithmetic in the first position

[29] "Intorno al 'Liber de numeris,'" p. 227. See above, pp. 59-60.

[30] Macrobius 1. 6. 77.

[31] Macrobius (1. 6. 80) does not divide the trunk but includes the *membrum virile* to bring the number to seven. Another writer, Anatolius, substitutes the neck.

[32] Probably the bright stars of Ursa Major, the dimmest of which is of third magnitude. Varro (Gellius 3. 10. 2) referred the number to the stars in both Ursa Major and Ursa Minor.

[33] Perhaps the best indications of this are the attention to arithmetic and the neglect of geometry in the writings of Martianus, Cassiodorus, Isidore, and Bede.

[34] For references on this subject see the article "Neopythagoreanism" in *The Oxford Classical Dictionary*. For Neopythagoreanism in Varro's writings see Leonardo Ferrero, *Storia del pitagorismo*, pp. 319-34.

among quadrivium studies in the curriculum,[35] finding its practical application in computing the dates of Easter and other movable feasts. *Computus* tables and treatises turn up with great frequency in medieval codices of scientific literature.[36] The mystical side of arithmetic —arithmology—especially intrigued the Fathers and writers like Cassiodorus, who saw fit to relate the numbers of the decad to the Bible and attached special numerological significance to the *two* testaments, the Holy *Trinity*, and the *five* books of Moses.[37] Such writers were responsible for generating in medieval Christian literature much involvement with the mystery of numbers.

Not to be overlooked as a factor in the dominance that arithmetic was gaining over the other quadrivium studies is the adoption of a new school manual. The preeminent position long held by Euclid's *Elements* in Greek mathematical studies was taken over by Nicomachus' *Introduction* in what meager attention was devoted to mathematics in the Latin schools. Nicomachus' book had aroused immediate interest when it appeared and within a few decades had been translated into Latin by Apuleius. Marrou attributes the shift in interest from geometry to arithmetic to this book.[38]

[35] Cassiodorus and Isidore inverted the traditional (Varronian) order, and placed arithmetic ahead of geometry. Friedmar Kühnert, *Allgemeinbildung und Fachbildung in der Antike*, pp. 50-70, discusses the order and arrangement of quadrivium subjects as they were treated by classical and post-classical authors.

[36] See Leonardi, "I codici" (1960), Index: *computo*, *computus*. For a bibliography on *computus* see L. Thorndike, "Computus," *Speculum*, XXIX (1954), 223-38; and for examples of *computus* see Clagett, *Greek Science*, pp. 161-65. The importance of *computus* studies is also evident from the many memorable Church controversies over the correct date for Easter—e.g., the Synod of Whitby, called by King Oswy in 663. At one time Oswy was fasting for Lent while the queen was feasting for the Resurrection. The controversies in Bede's day and the entire range of *computus* literature have been thoroughly studied by C. W. Jones in his edition of *Bedae Opera de temporibus*.

[37] Cassiodorus 2. 4. 8. Cf. Isidore 3. 4. 1-3. And for an elaborate discussion of number symbolism in secular and Christian literature in the Middle Ages see Curtius, pp. 501-9. For other applications of arithmetic see Fontaine, I, 344-45. Fontaine (I, 370-72) sees Isidore as the founder of the arithmological art in the Church.

[38] H. I. Marrou, *A History of Education in Antiquity*, p. 179. On the influence of Nicomachus' *Introduction* see chap. X of the D'Ooge, Robbins, and Karpinski volume.

A second translation of the work, by Boethius, gave the study of arithmetic new impetus. The opening arguments of Nicomachus' book appear to have shaped Boethius' career. Nicomachus gives fervent expression to the Pythagorean attitude, first found in Plato's *Republic*, that the path to philosophical truth lies in the mastery of the four mathematical sciences, and he goes on to argue that arithmetic is the foundation of all mathematical studies. Boethius, probably not yet twenty years of age at the time, began his career of philosophical study and writing by translating Nicomachus' *Introduction*, before he proceeded to the composition of his other mathematical manuals and to the study of philosophy. Boethius' translation of Nicomachus' *tessares methodoi* [four methods] as *quadruvium* (sic) is the earliest known instance of the use of the term.[39]

Martianus' extended section on arithmetic, forty-six pages long in the Dick edition, is one of the most important Latin expositions of Greek arithmetic from the early Middle Ages.[40] Although his ultimate sources were Nicomachus and Euclid, it is evident from a comparison of the three works that Martianus' immediate source was some compilation (or compilations) of the Nicomachean and Euclidean traditions, quite drastically revised during the intervening centuries. Martianus presents the definitions from Book VII of Euclid, with numerical examples, and the enunciations of twenty-five of the thirty-six propositions of Book IX and of the simpler ones from Book VIII, arranging

[39] Nicomachus 1. 4. 1-5. *Methodos* (literally, "a way after") means "pursuit of knowledge" or "mode of inquiry," "plan," or "system." *Quadruvium* (sic) means "a place where four roads or ways meet."

[40] Robbins (*Nicomachus*, tr. D'Ooge, pp. 138-42) presents a detailed comparison of the treatments given to the subject by Martianus, Cassiodorus, and Isidore, using Martianus as the basis of comparison because his treatment is the fullest. Robbins uses for his comparison with Isidore the chapters on arithmetic in Book III of the *Etymologies*, overlooking the more extended treatment, derived from Martianus, in Isidore's *Liber de numeris*. For a comparison of the *Liber* with Martianus see Leonardi, "Intorno al 'Liber de numeris,'" pp. 203-31. Bolgar (*The Classical Heritage*, p. 122), as is his wont, makes an arresting observation but offers no documentation: "The compiler of that *liber de numeris* which used to be attributed to Isidore, but is now regarded as almost certainly of Irish origin..." The book was in Braulio's list of Isidore's works, however. Fontaine, I, 373-82, treats the work as genuine and remarks at length on Isidore's borrowing from Martianus, his principal source.

them in a new order. Whereas Euclid always offers logical proofs, developed geometrically, Martianus gives arithmetical illustrations. Martianus' Nicomachean arithmetic includes some Euclidean material not found in Nicomachus.[41] Martianus takes up most of the topics of Nicomachean arithmetic but deviates from Nicomachus' order several times.

The section on arithmetic opens in the conventional way, with a definition of number (743): "A number is a collection of monads or a multitude proceeding from a monad and returning to it."[42] Martianus then classifies numbers into four types: even times even, odd times even, even times odd, and odd times odd.[43] Next follows a discussion of prime numbers (744), defined as "numbers that can be divided by no number (not divisible by the monad but composed of it)."[44] They are called prime because they arise from no number and are not divisible into two equal parts. Arising in themselves they beget other numbers from themselves; for even numbers are begotten from odd numbers, but an odd number cannot be begotten from even numbers. He concludes his remarks on prime numbers by saying that they must be considered beautiful (744).[45]

Next comes the arrangement of numbers into series (*versus*): first series, 1-9; second, 10-90; third, 100-900; fourth, 1,000-9,000. Some writers, he notes, include 10,000.[46] In the first series, the monad is not

[41] See *Nicomachus*, tr. D'Ooge, nn. to pp. 139-40.

[42] Cf. Euclid 7. Def. 2; Nicomachus 1. 7. 1; Boethius *De institutione arithmetica* 1. 3. On the various definitions given by Greek authors see *Nicomachus*, tr. D'Ooge, pp. 114-15; *Euclid*, tr. Heath, II, 280.

[43] Euclid 7. Defs. 8-10, omits odd times even. Nicomachus (1. 8. 3) subdivides even numbers only, into even times even, odd times even, and even times odd, as does Theon (ed. Hiller), p. 25. On the classification of numbers see Heath, *History*, I, 70-74; Karpinski in *Nicomachus*, tr. D'Ooge, pp. 48-50.

[44] Euclid Def. 11 defines a prime number as "measured by a unit alone." Cf. Nicomachus 1. 11. 2. On the definitions of prime numbers as given by Greek writers see *Nicomachus*, tr. D'Ooge, pp. 201-2; *Euclid*, tr. Heath, II, 284-85.

[45] Greek arithmeticians too could not rid themselves of the practice of giving epithets to numbers, calling some "friendly," some "perfect." See Heath, *History*, I, 74-76.

[46] E.g., Philo Judaeus *De plantatione Noe* 18. Elsewhere Philo designates the decimals 10, 100, and 1,000 as *kamptēres* [turning-points], a word of which Martianus' *versus* is a translation. Nicomachus does not give series a formal treatment,

a number[47] (to Geometry it is an indivisible point [746]).[48] The dyad is an even number; the triad is prime, "in both order and properties"; the tetrad belongs to even times even; the pentad is prime; the hexad, being odd times even or even times odd, is called perfect; the heptad is prime; the octad is even times even; the ennead is odd times odd; and the decad is even times odd. (745)

Arithmetic remarks that the only numbers that find favor with her are those counted on the fingers of both hands; beyond these digits contorted movements of the arms are required to encompass numbers represented by the lines and figures dealt with earlier by her sister Geometry.[49] For Arithmetic the beginning of the first series is the monad, for Geometry it is the point; numbers in the second series, beginning with the decad, are extended like a line; quadrate numbers (here representing surfaces) begin with 100, the first number of the third series; cubes (here representing solid figures) begin with 1,000, the first number of the fourth series. Arithmetic's representation of the decad as a line, the hecatontad as a square, and the milliad as a cube is not in accordance with the doctrines of Greek arithmeticians[50] and may be unique. This may be one of Martianus' own interpolations, signaled by a reminder of the presence of Arithmetic.[51] A moment

---

but he assumes (1. 16. 3) the reader's familiarity with the four series. On these series see F. E. Robbins, "Arithmetic in Philo Judaeus," *CP*, XXVI (1931), 349-50.

[47] Cf. Macrobius *Commentary* 2. 2. 8. On the monad as the beginning of numbers and not a number itself see *Nicomachus*, tr. D'Ooge, pp. 116-17.

[48] Nicomachus (2. 7. 1, 3) also notes the analogy between the geometrical point and the monad.

[49] The sole account of ancient Roman finger reckoning is found in chap. I of Bede's *De temporum ratione*. This account is frequently found as a separate tractate in medieval manuscripts. For a bibliography on ancient finger reckoning see *Bedae Opera de temporibus*, ed. Jones, pp. 329-30, and for a list of illuminated manuscripts showing the positions of the hands and body see Jones, ed., *Bedae Pseudepigrapha*, p. 54. See also Marrou, *History of Education*, pp. 157, 400-1.

[50] With Nicomachus (2. 7. 3) linear numbers begin with 2, plane numbers with 3 (vertices of a triangle), and (2. 13. 8) solid numbers (pyramid with triangular base) with 4. See Heath, *History*, I, 76-84; *Nicomachus*, tr. D'Ooge, pp. 54-60, 249-51.

[51] As noted above (pp. 32-33, 37), it usually happens, when we are made aware of a bridesmaid as the speaker, that Martianus introduces matter not found in conventional handbooks.

later (747) she speaks of 4 as the first square number and 9 as the second, and she points out that each successive odd number in a series progressing from the monad must produce a square number—a process that extends to infinity.[52]

After this interpolation Martianus returns to his classification of odd and even numbers, giving Euclid's definition of an even number (7. Def. 6) as one that is divisible into two equal parts, and the first part of Euclid's definition of an odd number (Def. 7) as one that cannot be divided into two equal parts.[53] He then observes that some odd numbers are merely uneven, like 3, 5, 7; while others (9, 15, 21) are also multiples of odd numbers; the Greeks classify the latter as odd times odd. Martianus gives the correct Greek forms for the terms even times even (ἀρτιάκις ἀρτιους), odd times even (περισσάκις ἀρτίους), and even times odd (ἀρτιάκις περισσούς), adding to the general impression that by his time the Nicomachean handbook tradition had not deteriorated as greatly as had the Euclidean tradition.

Next, and in proper order, comes the classification, with definitions, of numbers into (1) prime and incomposite, (2) composite in relation to themselves, (3) prime in relation to one another, and (4) composite in relation to one another. The first and smallest measure of all numbers is the unit. Numbers are susceptible of other measures, such as duplication or triplication. Some numbers have their sole measure in the unit; others, like 4 and 9, can be divided into other numbers; and still others, like 8, have more than one measure. Of numbers that are considered individually, those that have no measure but the unit are called prime and incomposite;[54] those that can also be measured by some other factor are called composite in relation to themselves.[55] Of two or more numbers taken together, those, like 3 and 4, that have no common measure except the unit are called prime to one another;[56]

[52] I.e., $1+3+5\ldots+(2n-1)=n^2$. Cf. Nicomachus 2. 9. 3. On square numbers in general see Heath, *History*, I, 77-79.

[53] The second part of Euclid's definition is "or one that differs by a unit from an even number."

[54] Cf. Euclid 7. Def. 11. Martianus is here dealing with products, not sums.

[55] Cf. *ibid.* 13.

[56] Cf. *ibid.* 12.

and those, like 9 and 12, that have some other common measure besides the unit, are called composite to one another.[57] (750-51)

Still another classification divides numbers into the perfect, the superabundant, and the deficient—the latter two being in Greek terminology "overperfect" (*hyperteleioi*) and "underperfect" (*hypoteleioi*).[58] A perfect number is one that is equal to the sum of its parts, like 6 and 28; a superabundant number is one the sum of whose parts is greater than the number itself, like 12; a deficient number, like 16, is one the sum of whose parts is less than the whole. (753)

Continuing to follow the style of a systematic teaching manual—that is, setting up his divisions of terms, defining them, and then elaborating[59]—Martianus next takes up plane and solid numbers. He finds that "the Greeks" call a number plane if it is the product of two numbers.[60] The factors of a plane number are represented as arranged along two sides of a right angle, resembling a *norma* [carpenter's square];[61] if one side is extended to a length of 4 and the other to 3, the product is 12, represented by the rectangle thus formed. According to the Greeks, he continues, a solid number is one which is the product of three numbers.[62] If above a surface representing 12, you place an identical quadrangular surface, a solid is produced, rep-

[57] Cf. *ibid.* 14. Martianus follows Euclid here. Nicomachus (2. 11) classifies odd numbers only, dividing them into (1) prime and incomposite, (2) secondary and composite, and (3) absolutely composite but relatively prime. See *Nicomachus*, tr. D'Ooge, p. 201.

[58] Euclid (7. Def. 22) defines only a perfect number. Nicomachus (1. 14. 3; 15. 1-2) and Theon (ed. Hiller, p. 45), define superabundant and deficient numbers as well. See *Euclid*, tr. Heath, II, 293-94.

[59] See above, pp. 38-39.

[60] It should be pointed out here that Nicomachus and Theon have a very different way of regarding plane and solid numbers from that of Euclid. For Euclid the unit is represented by a line of given length, and any linear number is then represented by the appropriate sum of unit lengths. A plane number is represented by a rectangle whose sides correspond to the two factors of the number; and so on. Nicomachus and Theon represent the unit not as a line of given length but as a point. A linear number is the sum of points arranged in one dimension; and so on. For further discussion of this view of numbers, see *Euclid*, tr. Heath, pp. 287-89.

[61] The Greek word for the instrument is *gnomon*. See Heath, *History*, I, 78.

[62] Martianus is again following Euclid Def. 17, but the examples he gives are his own or those of his immediate source.

resenting 24. Plane figures come from numbers arranged on a flat surface; solidity is produced when numbers are also arranged one above another. (754)

Martianus' discussion of surfaces is first Nicomachean (755), then Euclidean.[63] (756) The surface has numerous and varied forms, as illustrated by the figures which represent numbers. The first is a line; the triangle comes next after the line.[64] Surfaces with four angles either are square or have two sides that are longer than the smaller sides by one unit.[65] Polygons may be shown with sides of varying length. Among numbers that are elevated to solidity, the cube is seen to be perfection. To recapitulate, the smallest number that can be represented by a triangle is 3, by a square 4. The smallest number represented by a polygonal figure with an uneven number of sides[66] is 5. The smallest number represented by an oblong, with sides of unequal length, is 6.[67] The smallest solid number, representing a cube, is 8. (755)

Plane numbers and solid numbers are similar if their sides are proportional.[68] The plane numbers 6 and 600 are similar, since for 6 one side is 2, the other 3, and for 600 one side is 20, the other 30.[69] The solid numbers 24 and 96 are similar, one having sides of 4 and 3, making a plane surface of 12 and a solid of 24; the other having sides

[63] See above, n. 60.

[64] Cf. Nicomachus 2. 7. 3.

[65] The Greek term for the oblong figure is *heteromēkes* [heteromecic]. Cf. Nicomachus 2. 17. 1. Martianus' expression for heteromecic figures is the same as that used by Boethius (*De institutione arithmetica* 2. 26; Friedlein ed. 115. 9): *altera parte longiores.*

[66] Pentagonal figures. Cf. *ibid.* 2. 10, and see *Nicomachus*, tr. D'Ooge, pp. 243-44.

[67] On oblong numbers see Heath, *History*, I, 82-84.

[68] Now following Euclid Def. 21.

[69] It did not occur to Martianus to point out that 600 is also a solid number (2×3×100), if he was aware of the fact. Dick emended the text here and, in so doing, corrected Martianus' mistake and not the manuscripts, which read CC and CCC. That Martianus, and not the scribe who copied the archetype, was responsible for the mistake is evident from a statement below (761) in which Martianus says that 3 and 300 bear the superdimidius ratio to 2 and 200. Dick is doing the same thing here that Kopp chided Grotius for doing in an earlier passage (see above, p. 96).

of 8 and 6, a plane surface of 48 and a solid of 96. (756) In both illustrations Martianus is merely superimposing one plane surface upon another (effectively as though there were an altitude of 2) and is not dealing with solid numbers in a full and traditional way. Remigius, who understands the subject better than Martianus, points out the limited character of his treatment.[70]

Martianus next discusses relative number and the ratios of numbers (757), defining the terms which are written in Greek characters. Every number is a part of a larger number; the larger number is *produced* either by simple multiplication[71] or by a ratio of members[72] or of parts,[73] or by a combination of both multiplication and by ratio of members or of parts. Conversely a smaller number is *reduced* from a larger number by simple division[74] or by a ratio of members[75] or of parts,[76] or by a combination of these. These and the compound names for the sub-classes will be explained by Martianus in a moment. (759)

Then comes the relation that one number bears to another—that of equality, as in the case of 2:2, 3:3, or of a perfect number to the sum of its parts; or the relation of difference, when one number is greater,

[70] Remigius in a gloss on 384. 10 (ed. Lutz, II, 206): *Omnis crassitudo secundum alios huius artis auctores intus habet in medio quod intra exteriores superficies lateat. Ista vero crassitudo secundum Martianum nihil habet interius sed solas extrinsecus superficies.* For the traditional way of treating solid numbers in figures see Nicomachus 2. 17. 6 (tr. D'Ooge, pp. 256-57).

[71] The larger number is then called a "multiple" (*multiplicatus*). Cf. Nicomachus 1. 18. 1 (tr. D'Ooge, p. 214).

[72] *Ratio membrorum.* The larger number is then called by historians of mathematics "superparticular," a term given currency by Boethius, in his translation of Nicomachus. It contains within it the smaller number and one factor of it in addition, as 9 compared to 6. Cf. Nicomachus 1. 19. 1 (tr. D'Ooge, pp. 215 ff.).

[73] *Ratio partium.* The Boethian term for the larger number is "superpartient." Martianus defines the *ratio partium* below (759): "One number surpasses another number by a ratio of parts if the larger number contains within itself both the smaller number and some part or parts of it, as with 7 and 4." Cf. Nicomachus 1. 20. 1 (tr. D'Ooge, pp. 220 ff.).

[74] The smaller number is called by Boethius "submultiple" (*submultiplex*). Cf. Nicomachus 1. 18. 2; Boethius *De institutione arithmetica* 1. 23.

[75] The smaller number is called "subsuperparticular." Cf. Nicomachus 1. 19. 2; Boethius 1. 22.

[76] The smaller number is called "subsuperpartient." Cf. Nicomachus 1. 20. 3; Boethius 1. 28.

the other smaller. The latter relation occurs with a number that exceeds another in a ratio of members or of parts, or is exceeded by the other.[77] Numbers which bear the relation of equality are preferable to others, "for what can be better than an equal?" Though the difference between two numbers, greater and smaller, is the same, the ratio between those numbers is contrary. There is the same difference between 3 and 4 as between 4 and 3, but the ratio between those numbers is not the same.[78] (758) This will be explained later, he says, but the promise is not kept.

Following the customary procedure of handbook authors, Martianus now elaborates on his classification of the ratios of numbers: multiples (759-60), ratios of members (759, 761), and ratios of parts (762-63). In multiples there are the ratios of double, triple, quadruple, and beyond; a number is divided through the same steps, in reverse order: 4 is larger than 2 by the double ratio and 2 is smaller than 4 by the same. In ratios of members various names are given; when a larger number exceeds a smaller by half of the smaller (9:6), it bears the ratio of *superdimidius* (ἡμιόλιος) to the smaller; by a third, *supertertius* (ἐπίτριτος); by a quarter, *superquartus* (ἐπιτέταρτος); by a fifth, *superquintus*; by a sixth, *supersextus*; and so on. The reciprocal of *superdimidius* is called *subdimidius* (ὑφημιόλιος); of the *supertertius*, *subtertius* (ὑπότριτος); and of the *superquartus*, *subquartus* (πὑοτέταρτος); and so on.[79]

The remainder of Book VII is given over largely to numerical examples. Readers should not suppose that Martianus' absorption in simple arithmetic, at a level of the lower primary school grades, is necessarily characteristic of a medieval scientific mind. The same simple examples are found in Nicomachus; and Augustine, a precursor

[77] Nicomachus 1. 17. 1-5 takes up the relation of equality but does not discuss difference. Martianus has in mind what we express by the words "multiple of" and "factor of."

[78] On the differences being the same but the ratios being different cf. Nicomachus 2. 23. 1.

[79] Liddell and Scott's *A Greek-English Lexicon* (Oxford, 1940) cites ὑπότριτος ρυεϱποτέταρτος as found in Martianus only, and they may be Martianus' own forms, his translation of the terms back into Greek. P. Tannery, "Ad M. Capellae librum VII," *Revue de philologie*, XVI (1892), 137, suggests that they be emended to the regular forms ὑπεπίτριτον and ὑπεπιτέταρτον.

of the Middle Ages but surely not himself medieval, shows a similar absorption in the elementary arithmetic he derived from Varro.[80]

Simple examples are offered to show that the ratio of parts is closer to the supertertius in certain numbers and to the superquartus in others. When a larger number contains a smaller number and some third parts of it, the ratio is like the supertertius; or if some quarter parts of it, like the superquartus. For example, 5 exceeds 3, containing it and two third parts of it; 7 contains 4, and three quarter parts of it. But there is no ratio of parts that is *like* the superdimidius; for if a number contains another number and a half part of it, it *is* a superdimidius, as in the case of 6 and 4. (762)

Next Martianus gives numerical examples to illustrate the case of larger numbers being produced by a combination of multiplication and a ratio of members or of parts. Take 4 and 10; 10 is produced by the double and the superdimidius. Or take 4 and 14; 14 is produced by the triple and the superdimidius. In the case of 3 and 7, 7 is produced by the double and the supertertius. Or with 3 and 13, 13 is produced by the quadruple and the supertertius. And so on, with several more cases and examples.[81] (763)

Multiplication begins with the smallest ratio (double) and proceeds to larger and larger ones (triple, quadruple, etc.), one exceeding another by ratio of members or of parts. But the ratio of members begins with the superdimidius and proceeds to the supertertius, superquartus, and to even smaller ratios. In the ratios, terms that are minimal are called *pythmenes* [root-forms][82] by the Greeks. The minimal terms of the superdimidius are 2 and 3, of the supertertius 3 and 4, of the superquartus 4 and 5. (764-66)

At this point Martianus introduces an attractive little digression (767)—the source is unknown, but the observation is obviously not his own—on the sequence of the discoveries of the ratios: Multiples were

[80] *De civitate Dei* 1. 30; 18. 23. Readers may wish to skip over the remainder of Martianus' account, which makes dull reading but is included here for the sake of completeness.

[81] Cf. Nicomachus 1. 22. 3-6.

[82] Cf. *ibid.* 1. 19. 6; 1. 21. Martianus' text is corrupt here. Tannery, "Ad M.C. librum VII," pp. 137-38, offers an emendation, πυθμένας *pythagoricus Thymaridas nominabat*, which Dick does not note in his apparatus.

probably discovered first, then ratios of members, then parts. The ratios of double, triple, quadruple, and so on, presented no complexities. It is natural to suppose that the superdimidius proceeded from the double ratio, the supertertius from the triple, and the superquartus from the quadruple. For a person comprehending the double is beginning to comprehend the dimidius (as 4 is the double of 2, so 2 is the half, or dimidius, of 4). The superdimidius was discovered by adding 2 again to 4, the sum of 2 and 2. The number 6 was then seen to result from the tripling of 2; and in adding 2 more, the ratio of supertertius was discovered. Finally, when numbers arose that did not fit into the regular ratios, questions naturally were posed about how many parts of one number were found in another—"in order thus to establish some relationship of one number to another."

Ratios of numbers concluded, Martianus now returns to numbers *per se*, discussing evens and odds first. An even number, in any multiple of itself, remains even: 2, 4, 8, 16 (doubling); 1,[83] 4, 16, 64, 256 (quadrupling). But an odd number multiplied by an even number reverts to an even number.[84] An odd number multiplied by an odd number remains odd.[85] The explanation is that, whether there is an even or an odd number of even numbers to be added, the sum is even.[86] But an odd number of odd numbers to be added gives only an odd sum.[87] If evens are added to evens, the sum is even.[88] If odds are added to odds, the sum is also even. If a number of either class (odd or even) is subtracted from an even number, a number of that class remains;[89] but the opposite is true of an odd number: if an even

[83] The Pythagorean view of the monad as both even and odd was widely adopted by popular writers. Cf. Macrobius 1. 6. 7; Theon, ed. Hiller, p. 22; Calcidius 38. And see Heath, *History*, I, 71, for possible explanations. Martianus has in mind square (plane) numbers originating from the monad (a point).

[84] From here to the end of the book Martianus follows Euclid's propositions, but not his proofs. Euclid develops his proofs by lines; Martianus illustrates by numbers. Cf. Euclid 9. 28.

[85] Cf. *ibid.* 29.

[86] Cf. *ibid.* 21-22.

[87] Cf. *ibid.* 23.

[88] Cf. *ibid.* 21.

[89] Cf. *ibid.* 25.

number is taken from an odd number, the remainder is odd;[90] if an odd number is taken from an odd, the remainder is even.[91] (768-70)

Any number that has an even half is an even times even number; likewise any number that is doubled, beginning with 2, or any number that is produced by quadrupling, octupling, and such belongs to the class of even times even.[92] Any number that has an odd half is even times odd.[93] If any number neither arises from 2 by doubling nor has a half that is odd, it belongs to the even times even class, but it originates in the class of even times odd.[94] Consider the number 12. It neither arises by doubling from two nor has an odd half; but it arises by duplication, from 6, a number that belongs to the even times odd class. (771)

Martianus has already classified numbers into prime or composite. He now elaborates on these classes and provides numerical examples of each. All prime and incomposite numbers are odd and have no factor but the unit.[95] (773) Even numbers are composite in relation to themselves, whether they come from evens or odds. Odd numbers may also be composite in relation to themselves, if they are the product of odd numbers. (772) No two even numbers are prime to one another, because they have some common measure (in duplication). If a number that is prime and incomposite is taken with another number that is composite in relation to itself, the two are found to be prime to one another [!] It does not matter if one number is measured by some other part than the unit if this is not true of the other (774). Or take two or more numbers that are composite in relation to themselves and also in relation to each other; the addition of an incomposite number to the group causes them to become prime to one another (3 included with 4, 6, and 8). Two numbers that are composite in relation to themselves (9 and 25, or 8 and 9), can, when brought together, be

[90] Cf. *ibid.* 27.

[91] Cf. *ibid.* 26.

[92] Cf. *ibid.* 32.

[93] Cf. *ibid.* 33.

[94] Cf. *ibid.* 34.

[95] Cf. Euclid 7. Def. 11; Nicomachus 1. 11. 2. The number 2 satisfies Euclid's definition of a prime number and is regarded as prime by Aristotle, but in Nicomachus' treatment all prime numbers belong to the class of odd numbers. See Nichomachus, tr. D'Ooge, p. 201, n. 2.

prime in relation to each other. (775) In certain cases odds and evens, like 9 and 12, are composite to each other, since both arise from triplication. Noteworthy is the fact that no even number from evens, only evens from odds, can be composite with an odd number (776): 9 is not composite with 4, 8, 16, but is composite with 12, 24. Not every odd number that is composite in relation to itself can be composite with every number that comes from odds, because the two numbers may not be divisible by the same measure. Thus 9 and 50 are not composite, because 50 does not arise from triplication (777).

Then follows the rapid enunciation of many propositions from Euclid's *Elements*. (778-801) The translated statements are wordier than the original and are not in Euclid's order. In place of Euclid's proofs, Martianus offers numerical illustrations.

If either of two numbers that are prime to each other is composite in relation to itself, a measure of that number is not composite with the other number (Euclid. 7. 23). If two numbers are prime to each other and one of them multiplies itself, the product will be composite with the other number.[96] If two numbers that are prime to each other multiply themselves, the products will be prime to each other (7. 27). And if two numbers are prime to each other and one of them multiplies itself, and if that number multiplies the product again, the resulting number will not be composite with that other number (7. 27). If two numbers are prime to each other and if each one multiplies itself, and multiplies the product again, the resulting numbers will also be prime to each other (7. 27). If two numbers that are prime to each other are added, the sum of the two numbers cannot be composite with either of the former numbers (7. 28). If two numbers are taken together with a third and all are prime to one another, the product of the two numbers cannot be composite with the third number (7. 24). If numbers do not contain some part of a number, they cannot be composite with it (7. 29). If three numbers joined together[97] are the least of those which have the same ratio with them, any two of these added together are not composite with the third

[96] This statement of Martianus' is an erroneous transmission of Euclid 7. 25 ("If two numbers are prime to one another, the product of one of them into itself will be prime to the remaining one").

[97] I.e., in continued proportion.

(9. 15). If an odd number is not composite with another number, it will not be composite with the double of that number (9.13). Given two pairs of numbers in which neither member of the first pair is composite with either member of the second, the product of the first pair cannot be composite with either number of the second pair (7. 26). The least numbers of those which have the same ratio with them are prime to each other (7. 22). Any number is either prime and incomposite or, if it is composite in relation to itself, is measured by some prime number (7. 32, 31).[98]

Next comes (785-90) the statement of some problems from Book VII of the *Elements*. The procedures are explained arithmetically. If two numbers, a greater and a smaller, are composite with each other, find their largest and their smallest common measure (Euclid 7. 2). Of three numbers which are composite with one another, find their largest and their smallest common measures (7. 3). Given two numbers, find the smallest number which they measure (7. 34). Given three numbers, find the smallest number which they measure (7. 36).

Martianus concludes the section on arithmetic with statements of various Euclidean propositions, with an excessive use of numerical examples. (791-801) Given two pairs of numbers, larger and smaller, of such sort that there is the same ratio between the larger and smaller pairs, as often as the larger measures the larger, the smaller will measure the smaller (7. 20).[99] If the unit measures any number as many times as another number measures a fourth number, it will happen that, as many times as the unit measures the first number of the second pair, the number which had its measure in the unit will measure the last number of the second pair (7. 15).[100] If two numbers are multiplied and some prime and incomposite number measures the product, it must also measure both of the original numbers.[101] Let as many numbers as you wish be placed in increasing order in a continued propor-

[98] This paragraph is a summary of the contents of Martianus 778-85.

[99] I.e., if $a:b=x:y$, and $a<x$, then $a:x=b:y$.

[100] I.e., if $1:x=y:z$, then $1:y=x:z$.

[101] Euclid (7. 30) says one of the original numbers, Martianus says both; and the numbers he uses to illustrate (8 and 10) both have a measure in 2. But take 7 and 12; two measures the product (84) but does not measure 7. However, it does measure both of two other factors (6 and 14) of 84.

tion; if the first number measures the last, it measures the second as well, and all the others following it; if it measures the second, it also measures the last and the intervening ones; if finally it measures any one, it measures all (8. 7). If as many numbers as you please, beginning with the unit, are in a continued proportion, as many prime numbers as measure the last number will measure the number which is next to the unit (9. 12). If as many numbers as you please, beginning with the unit, are in a continued proportion, the smaller always measures the larger by some one of the other numbers that are in the same proportion (9.11). If as many numbers as you please, beginning with the unit, are in a continued proportion, and the number next to the unit is prime, the greatest number will not be measured by any except those that are in the same proportion (9. 13). If two prime numbers measure the least number, no other prime numbers will measure it (9. 14). If a square number measures another square number, the side of the first square will also be the measure of the side of the other (8. 14); and if a square number does not measure a square number, the side of one will not measure the side of the other (8. 16). If a cube number measures another cube number, the side of the first cube will also measure the side of the other (8. 15); and if one cube number does not measure another cube number, the side of the first will not measure the side of the other (8. 17). Any number that is measured by another number gets the name of the measure from the same number that makes the measure (7.37). If a number has a part, it will be measured by a number that has the same name as the part (7. 38).

Arithmetic brings her discourse to an end abruptly with a short poem and apologizes, as did the other bridesmaids, for being tedious and prolix:

> Time warns me to bring my discourse to a close.
> Should boredom steal upon the heavenly throng
> I, old "Number-keeper," would be driven from the sky . . .

Her suggestion that a more extended treatment of the subject "would befit Attic sages" confirms the impression left by the whole of Martianus' presentation of the discipline: that he appropriated here, or relied heavily upon, some Latin *ars arithmetica* and that he did not use Greek sources.

As Arithmetic seats herself, a reverential silence comes over the august company of gods and philosophers. Pythagoras, with all his disciples, and Plato, the while expounding esoteric doctrines from his *Timaeus*,[102] venerate the lady with mystical words of praise. Mercury is a particularly pathetic figure as he sits in rapt admiration of Arithmetic's performance. She is to him the most erudite and eloquent of all the bridesmaids. (803) All the spirit and subtlety of the Mercury of Plautus and Lucian have vanished, and he is shortly to get the sort of mate he deserves–a drab personification of antique handbook learning, an insipid cabbage that the Roman satirist Juvenal (*c.* A.D. 120) had already found warmed over by the masters of the rhetorical schools of his day.[103]

[102] For as long as Latin science was able to maintain a position of respect in Western Europe, the *Timaeus* was its bible (see above, p. 11). (The *Timaeus* is the book in Plato's hand in Raphael's painting of the ancient philosophers, "The School of Athens.") Aristotelian tenets are widely disseminated in Latin cosmographic traditions of the first Christian millennium, but the basic structure and conceptions of cosmography in this period bear more Platonic than Aristotelian characteristics. A new, definitive edition of Calcidius by J. H. Waszink appeared in 1962; Waszink places Calcidius at the close of the fourth century–without much assurance. Previously he was generally placed in the first half of the fourth century.

[103] Juvenal *Satires* 7. 150-54.

On Astronomy

THE SILENCE following the acclamation for Arithmetic's discourse is broken by a scene of rowdy brawling, a travesty of the refined imagery and precious banter in a scene by Vergil which may have inspired this one.[1] Silenus, his veins swollen with an overdraught of wine, has been snoring through Arithmetic's lecture. Suddenly he emits a thunderous belch. The entire party is convulsed with laughter, and pandemonium erupts. The attendants of Venus and Bacchus take over and the wine flows. Saucy Cupid runs up to Silenus and gives his ruddy bald head a resounding clap of the palm. The besotted old man slowly awakens and, peering through bloodshot eyes, staggers to his feet. He sways and reels about, then slumps to the floor. Satyr, on orders from Bacchus, heaves Silenus to his shoulders and drapes his bloated body, like a wineskin, about his neck.

At this moment Martianus is sharply reprimanded by Satire for introducing a brawl into an august senate of the gods when Astronomy is about to discourse on "the hallowed planets," "yon Herdsman Boötes," "brilliant Canopus," "the blazing horns of the ever-changing moon," and "the slanting belt of the zodiac," all plainly visible from the gods' vantage in the canopy of the heavens. Martianus responds to Satire's abuse with a question: "Am I to eschew all creatures of the imagination and not relieve the boredom of my readers with some mirth and drollery? Come to your senses, Satire, leave off your tragic ranting, and take a hint from the young Pelignian poet: Young lady, take my advice and smile."[2] Apollo steps out to usher in the next bridesmaid:

Before their eyes a vision appeared, a hollow ball of heavenly light, filled with

[1] *Eclogues* 6. 14-26.

[2] The verse is quoted from Martial (*Epigrams* 2. 41), who plainly states in the next line that he thinks the Pelignian poet (Ovid) offered this advice. Martianus' ascribing the quotation to the earlier source and suppressing the intermediary's name is in keeping with the general practice of late Latin and medieval writers.

transparent fire, gently rotating, and enclosing a maiden within. Several planetary deities, ominous or propitious, were bathed in its glare, the mystery of their behavior and orbits revealed. Even the fabric of the celestial sphere shone forth in the flashing light. Lesser deities, ethereal, terrestrial, marine, and subterrestrial, were astounded at the miraculous sight . . . and offered the maiden a seat of honor. Decked with gems and decorously arrayed, she stepped forth nimbly from the sphere. Her brow was starlike and her locks sparkled. The plumage on her wings was crystalline, and as she glided through the sky it took on a golden hue. In one hand she held a forked sextant,[3] in the other a book containing calculations of the orbits of the planets and their forward and retrograde motions. These were delineated in metals of various colors [810-11].

Astronomy, like her sister Geometry, is a peregrinator of the universe. She has traversed all the heavens and can reveal the constellations lying beneath the celestial antarctic circle. There is no reason to suppose here that Martianus, as a North African, was more familiar with southern skies than classical astronomers and geographers. He was strictly a handbook compiler, using data passed down from classical and Hellenistic Greek sources. Moreover, Carthage, the city of which he calls himself a fosterling, is actually slightly to the north of Rhodes, the home of Posidonius, the Hellenistic Greek who figured most prominently in the popular traditions of Latin cosmography.[4]

Astronomy tells us that she is also familiar with the occult lore of Egyptian priests, knowledge hoarded in their sanctums; she kept herself in seclusion in Egypt for nearly forty thousand years, not wishing to divulge those secrets. She is also familiar with antediluvian Athens.[5] She knows that she might have excused herself from discoursing at this time by referring the wedding guests to the astronomical books of Eratosthenes, Ptolemy, and Hipparchus; but she feels a sense of

[3] 811 (Dick 429. 3-4): *cubitalem fulgentemque mensuram*—probably not to be translated as a "measuring rod, a cubit in length." Émile Mâle (pp. 79, 85), translating it as "a bent instrument . . . which serves in measuring the altitude of stars," had in mind the interpretation given to *cubitalem* by medieval illustrators and painters. Remigius glosses both *radius* (see above, p. 125) and *mensura* as *virga geometricalis* (ed. Lutz, II, 130, 248).

[4] See Stahl, *Roman Science*, chap. IV.

[5] The story of Athens before the flood—an earlier flood than that of Deucalion, according to Proclus—comes from Plato's *Timaeus* 23c. On the Egyptian sojourns of Astronomy and other bridesmaids and on the restoration of the liberal arts after the flood, see Lutz, "Remigius' Ideas on the Origin of the Seven Liberal Arts." pp. 34-39.

obligation toward Mercury, and since his bride Philology also wants to hear her, she will begin.

First it should be pointed out that Astronomy's reference to the major Greek astronomers was a deception practiced by nearly all the Latin handbook authorities of the Empire and early Middle Ages. Little was known about these Greek writers beyond the awesomeness of their reputation. Hipparchus' greatest discovery, that of the precession of the equinoxes, is not even mentioned by the Latin writers. Eratosthenes' brilliant method of calculating the circumference of the globe, and the figure he obtained (252,000 stadia), was known in the West only as an isolated datum. Ptolemy's *Geography* was read and used to good effect by Ammianus Marcellinus in his *Histories*, but Ammianus was a Syrian Greek, not Latin. That Boethius translated some Ptolemaic work on astronomy seems likely from contemporary and later testimonies to such a translation and a title *De astrologia*. The translation, if it was made, was not of the *Almagest*, but of a shorter manual by Ptolemy or a handbook in the Ptolemaic tradition. The solution to this vexing question appears to await any scholar who will avail himself of a discovery made by Professor Ullman shortly before his death.[6] Despite the obvious fraudulence of Martianus' citations

[6] First let us review the evidence for a Boethian translation of a book on astronomy by Ptolemy. Boethius expressed the intention, in the Preface of his first work, *De arithmetica*, to compose treatises on each of the mathematical disciplines. Next we have an ambiguous and flattering letter (*Variae* I. 45), addressed by Cassiodorus (in Theodoric's name) to Boethius, when Boethius was about twenty-seven years of age, and stating that Pythagoras the musician, Ptolemy the astronomer, Nicomachus the arithmetician, Euclid the geometrician, Plato the theologian, Aristotle the logician, and even Archimedes the mechanician had become available in Latin, thanks to Boethius. This letter does not indicate, as has sometimes been supposed, that Boethius translated Ptolemy's *Almagest* and the mechanical treatises of Archimedes. Cassiodorus was obviously impressed with Boethius' talents and was confusing expressed intentions with actual accomplishments. Lastly there are two letters of Gerbert (Nos. 8 and 130 [Migne, *PL*, CLXIII, cols. 203, 233]): in the first, he refers to a Bobbio manuscript containing eight books of Boethius *De astrologia*; in the second, he asks a monk at Bobbio to make copies of M. Manlius *De astrologia*, Victorinus *De rhetorica*, and Demosthenes *Ophthalmicus*. Manlius has generally been identified as Manlius Boethius, and is so taken by a recent translator (Harriet Pratt Lattin, *The Letters of Gerbert*, p. 169); but R. Ellis (ed., *Noctes Manilianae* [Oxford, 1891], pp. 229-30) and A. E.

scholars continue to credit them, and one scholar recently took the trouble to point out the discrepancies between Martianus and Ptolemy.[7]

Astronomy's discourse, which occupies only thirty-nine pages in the Teubner edition by Dick, is a fragment that breaks off abruptly. The extant portion takes up all the conventional topics of an *ars astronomiae*, and the missing portion probably contained little more than a closing setting scene, such as is found in all the other discipline books. As Book VIII stands, it is the shortest book in the entire work. It is also the best of Martianus' quadrivium books and, considering its small compass, the best-proportioned and most satisfactory treatise on astronomy in the extant Latin literature before the Greco-Arabic revival. We can understand its eminence among manuals of the mathematical disciplines if we note that astronomy was the most highly developed of the Greek sciences[8] and if we accept the surmise, sug-

---

Housman (ed., *M. Manilii Astronomicon liber primus*, pp. lxix, lxxii) identify him as Manilius, author of the *Astronomica* (*c.* A.D. 15), a poem on astrology; and H. W. Garrod (ed., *Manilii Astronomicon liber II*, pp. lix-lxi) inclines to agree. Leonardi ("Nuove voci poetiche," pp. 162-63), thinks that the work on astrology referred to in Gerbert's Letter 8 was by Manilius and that Gerbert mistook it for a work by Boethius. More recently B. L. Ullman (p. 278) found all three works referred to in Letter 130 listed successively in the tenth-century Bobbio catalogue (Becker, No. 32) and he examined the manuscript containing the astronomical work attributed to Boethius. He found that about one-fifth of the manuscript is devoted to an *ars geometriae*, another fifth to gromatic excerpts, and almost half to astrological material. This manuscript he took to be the copy requested by Gerbert in Letter 130, and a copy of the manuscript of Boethius *De astrologia* in eight books, described by Gerbert in Letter 8. The reference to eight books has always puzzled scholars and led some to conclude that Boethius' work was a long one, but Ullman points out that the *ars geometriae* alone occupies five books. Unfortunately Professor Ullman either did not have an opportunity or did not see fit to examine the astrological portion of the manuscript to determine the Manilian, Boethian, or Ptolemaic character of its contents.

[7] See above, p. 141, n. 50.

[8] Derek J. de Solla Price, *Science Since Babylon*, pp. 6-8, assigns a unique importance to Greek mathematical planetary theory and says that "in all the branches of science in all the other cultures there is nothing to match this early arrival of a refined and advanced corpus of entirely mathematical explanation of nature."

gested earlier,[9] that the professional character of Martianus' treatise reflects a sound and well-preserved Greek tradition which may stem from a Varronian handbook derived from Posidonius.

Moreover, Book VIII was the most popular of Martianus' quadrivium books in the Middle Age[10] and established its author as one of the leading authorities on astronomy. Martianus' principal rivals, Calcidius and Macrobius, were popular largely because they expounded Neoplatonic cosmography. Calcidius' astronomy was too abstruse to be appreciated before the late Middle Ages. His lengthy section on astronomy in his commentary on Plato's *Timaeus* is a technical treatment involving mathematics that would not have been comprehensible to Martianus; for, as T. H. Martin discovered in 1849,[11] it is actually a largely free, occasionally literal, translation of part of a *Timaeus* commentary compiled by Theon of Smyrna.[12] The cosmographical section of Macrobius' *Commentary*, like Martianus' Book VIII, was circulated as a separate treatise on astronomy. Its popularity may be ascribed to its simplicity and clarity of exposition and to its overtones of Neoplatonic fervor.

It is not surprising that Book VIII has been of greatest interest to historians of science. As we shall see, their interest has been stimulated by Copernicus, who singles out Martianus to bestow lavish praise upon him for propounding a theory of heliocentric orbits for Venus and Mercury—though Copernicus was aware, as he says, that other Latin writers held the same view.[13] The theory, on good authority ascribed to Heraclides of Pontus (*c.* 340 B.C.), became a regular feature of popular handbooks in antiquity. Copernicus could have referred to a fuller account of Heraclides' theory. In fact he was also aware that Aristarchus of Samos (*c.* 260 B.C.) went a step further than Heraclides

[9] See above, pp. 51-52.

[10] See above, p. 71.

[11] *Theonis Smyrnaei Platonici liber de astronomia*, ed. T. H. Martin (Paris, 1849), p. 18.

[12] E. Hiller, "De Adrasti Peripatetici in Platonis Timaeum commentario," *Rheinisches Museum*, new ser., XXVI (1871), 582-89, argues cogently that Calcidius translated from Adrastus' commentary, not Theon's. If Hiller's argument is valid —and most scholars accept it—Theon becomes a plagiarist, since his text is so close to Calcidius'.

[13] Copernicus *De revolutionibus orbium coelestium* 1.10.

to propound heliocentric orbits for all the planets, including the earth, and make the moon a satellite of the earth.[14]

Astronomy continues her grandiloquent exordium: The universe is spherical in shape, composed of four, and only four, elements (*ex quattuor elementis isdem totis*). The earth is stationary, at the center and bottom of the universe.[15] The softness of rarefied bodies is surrounded by its very condensations[16] into certain set paths and intervals of circles (*subtilium corporum teneritudinem suis coactibus circumdatam in quasdam sectas vias et circulorum intercapedines*). The natures of these bodies, coalescing by their own surgings, are diffused the entire way round in globular belts and circles (*suis fluctibus adhaerentes naturas undiquesecus globoso ambitu orbibusque diffundi*) . . . If every belt of the encompassing substances is found to be homogeneous, no circles can waver from their ethereal tracts (*si igitur sui*

[14] This appears to have been deliberately suppressed by Copernicus. A footnote in the scholarly Thorn edition, which commemorated the fourth centenary of Copernicus' birth (*Nicolai Copernici Thorunensis De revolutionibus orbium coelestium libri VI* [Thorn, 1873], p. 34), contains a statement deleted by Copernicus from the final draft of his manuscript: *Credibile est hisce simi libusque causis Philolaum mobilitatem terrae sensisse, quod etiam nonnulli Aristarchum Samium ferunt in eadem fuisse sententia.* Then in the dedicatory Preface, in which he tells of his search through the classical literature for precursory statements of heliocentricism, he quotes a passage from [pseudo-] Plutarch about the views of Heraclides and Ecphantus but omits to mention the clear statement a few pages earlier about the heliocentric views of Aristarchus. See Heath, *Aristarchus*, p. 301, who credits Gomperz with pointing out the footnote in the Thorn edition; and Angus Armitage, *Copernicus: The Founder of Modern Astronomy* (London, 1938), pp. 87-90. Rudolf von Erhardt and Erika von Erhardt-Siebold (*Isis*, XXXIII [1942], 599-600) argued the almost certain acquaintance of Copernicus with Archimedes' *Arenarius*, the work that contains the most authoritative and best account of Aristarchus' theory. On the deletion by Copernicus of the passage about Aristarchus see T. W. Africa, "Copernicus' Relation to Aristarchus and Pythagoras," *Isis*, LII (1961), 406-7.

[15] Ever since Aristotle (*De caelo* 2. 13-14) set forth his doctrines and proofs of the natural places of the four elements—earth, being the heaviest of the four elements, is at the bottom of the universe, and objects falling from all directions upon the earth's surface fall toward the center—the notion that the earth was at the center (= bottom) of the universe became a commonplace in cosmographic literature. Cf. Pliny 2. 11; Cleomedes 1. 11; Macrobius *Commentary* 1. 22. 1.

[16] The meaning given to *coactibus* by the *TLL*.

similis omnis circumagentium naturarum ambitus reperitur, nulli possunt aetherium tractum circuli variare). When we use the word "circles," we do not intend to convey a notion of corporeal demarcations of a fluid substance; we are merely illustrating the risings and settings of planetary bodies as they appear to us (*nos igitur circulos non ita dicemus, ut liquentis naturae discrimina corpulenta fingamus, sed ut ascensus descensusque ad nos errantium demonstremus* [814-15]).

Astronomy is piqued by mere mortals who try to represent the celestial axis and poles on an armillary sphere.[17] The poles, she explains, protrude from the hollows of the perforated outer sphere, and openings and pivots have to be imagined—something that you may be sure could not happen in a rarefied and supramundane atmosphere (*cum poli velut perforatae exterioris sphaerae cavernis emineant, et hiatus quidam cardinesque fingantur, quod utique subtilibus aethereisque accidere non potuisse compertum*). Her reference to axis, poles, and celestial circles must be understood in a theoretical sense, as distinctions applied not to transitory conditions of the heavens but to calculations of intervals (*sicubi igitur intelligentiae edissertandique proposito vel axem vel polos vel circulos perhibebo, ideali quadam prudentia, non diversitate caeli discreta, sed spatiorum rationibus* [815-16]).

Such is the flamboyant jargon of Martianus' bridesmaids, of some interest to philologists but of none to historians of science. The audience duly awed, Martianus is now ready to present sober handbook materials. The opening is a conventional one, listing and defining celestial parallels and circles.[18] There are in all ten great circles, five of them called "parallels," or by Latin writers, "equidistants." First he defines the parallels, which gird the sky latitudinally; then the colures, which are longitudinal; then the oblique circles (zodiac and Milky Way); and lastly the horizon. (817-26)

[17] Varro points to another limitation in the use of armillary spheres in a passage quoted by Aulus Gellius (*Attic Nights* 3. 10. 3). There is verbal similarity here, and either Varro or Gellius probably inspired Martianus' thought.

[18] Cf. Theon (ed. Hiller), pp. 129-33; Cleomedes 1. 2. 11-12; Geminus 5; Calcidius 65-68; Macrobius *Commentary* 1. 15. 12-18; *Isidore* 3. 44-46.

His definitions of the parallels are standard ones:[19] The parallels have the same poles as the universe. The arctic circle, at its lowest point, just grazes the northern horizon, the constellations within it always being visible; the antarctic circle, at its highest point, grazes the southern horizon, the constellations within it always invisible. Our reasoning powers indicate it to be of the same extent as the arctic circle. The two tropics mark the northern and southern limits reached by the sun at the summer and winter solstices. The celestial equator, equidistant between the tropics, marks the equal length of day and night. (817-22)

The two colures are longitudinal celestial circles, passing through the poles at right angles to each other and cutting the girth of the universe into four equal segments. Some authorities prefer to trace them by beginning at the north or south pole. Martianus says he prefers to follow the authority of Hipparchus, who traces them from the equinoctial and solstitial points in the ecliptic, precisely at the eighth degree of Aries and Cancer.[20] (823-24)

Of the two oblique circles the zodiac is tangent to the celestial tropics of Cancer and Capricorn; it bisects the celestial equator, but the angles of intersection are not equal. It is marked off into twelve segments and furnishes a path for the sun, the moon, and the five planets.[21] (825) Regarding the Milky Way Martianus naïvely remarks that it plainly has a much greater girth than the other celestial circles, since it rises on the borders of the arctic circle and sets on the horizon of the antarctic circle, and appears to traverse nearly the entire celestial sphere. He scorns those writers who refuse to include it among

[10] The correspondences with Geminus (5. 1-9) are close. Cf. Theon (ed. Hiller), pp. 129-30; Cleomedes 1. 2. 11-12; Macrobius *Commentary* 1. 15. 13.

[20] Cf. Geminus 5. 49-50. Theon (ed. Hiller), p. 132, says that according to some writers the meridian is called a colure. Macrobius (1. 15. 14) has the colures intersect at the north pole and cut the zodiac at the solstitial and equinoctial points, but he does not believe that they extend to the south pole. The tradition assigning to Hipparchus the location of the vernal point at Aries 8° may have been correct. Otto Neugebauer, *The Exact Sciences in Antiquity*, p. 188, points out that the use of Aries 8° as the vernal point appeared in Greece about the time of Hipparchus. Eudoxus and earlier writers used Aries 15°.

[21] Cf. Geminus 5. 51-53; Theon (ed. Hiller), p. 130; Cleomedes 1. 14. 18; Macrobius *Commentary* 1. 15. 8-12.

the celestial circles.[22] The celestial horizon, the demarcation of the upper and lower worlds, concludes Martianus' list of celestial circles.[23] (826)

Astronomy relates that she herself has drawn the celestial circles in the sky with her compasses. She has set a brilliant star at the celestial pole and about it has drawn the arctic circle, tracing it from the head of Draco through the right foot of Hercules, the middle of the breast of Cepheus, the front paws of Ursa Major, and back to Draco's head.[24] (827) The classical Greek astronomers understood that the location of the arctic circle would vary with the latitude of the observer; Geminus points out that a circle traced through the front paws of Ursa Major applies to observers on the island of Rhodes.[25]

Again setting a compass point on the polestar she has traced the summer tropic through the eighth degree of Cancer, the chest and belly of Leo, the shoulders of Serpentarius, the head of Cygnus, the hoofs of Equus, the right hand of Andromeda, the left shin and left shoulder of Perseus, the knees of Auriga, the head of Gemini and back to the eighth degree of Cancer.[26] (828)

The celestial equator, equidistant from the poles, is traced from the eighth degree of Aries, to the retracted hoof of Taurus, thence to the belt of Orion, through the elevated coil of Hydra, through Crater and Corvus to the eighth degree of Libra; then it passes to the knees of

[22] Handbook authors do not agree in their lists of celestial circles. Some count eleven circles, some ten, and some fewer than ten. Theon and Cleomedes omit the Milky Way. Geminus (5. 11) includes it, calling it the only visible circle, the others being theoretical. Macrobius (1. 15. 2-7) gives the Milky Way the most elaborate treatment because it is the meeting place in heaven for deserving souls in *The Dream of Scipio*. Martianus omits the meridian from his list.

[23] Cf. Theon (ed. Hiller), p. 131. Geminus (5. 56) distinguishes between the theoretical and visible horizons. Macrobius (*Commentary* 1. 15. 17-18) ineptly includes the visible horizon in his list of celestial circles and omits the theoretical horizon.

[24] Most handbook authors do not trace the celestial circles through the constellations. However, Aratus' *Phaenomena*, the bible of popular astronomy, does trace the two tropics and the celestial equator in this way.

[25] Geminus 5. 3. Geminus and his teacher Posidonius lived in Rhodes.

[26] Martianus' tracing of the Tropic of Cancer is the same as that found in Aratus 480-96.

Serpentarius, through Aquila to the head of Pegasus, and back to the eighth degree of Aries.[27] (829)

The winter tropic begins at the eighth degree of Capricorn, passes through the feet of Aquarius and the end of the tail of Cetus, thence to Lepus and the front paws of Canis Major; it goes through Argo and the back of Centaurus to the sting of Scorpio, then through the extremity of Sagittarius' bow, and back to the eighth degree of Capricorn.[28] Beyond the winter tropic is the celestial antarctic circle. Astronomy says she can trace this through its constellations, too, for no part of the heavens is unfamiliar to her; however, she prefers not to disclose phenomena that are not verifiable by observers in the northern hemisphere. (830-31)

Tracing the colures through the constellations is less common, because they pass beneath the horizon in the southern hemisphere. The equinoctial colure begins at the equinoctial point, the eighth degree of Aries, grazes the far angle of Triangulum, touches the top of Perseus' head and his right arm, and cuts through his hand; crossing the arctic circle to the celestial north pole, it then passes through the tail of Draco to the left side of Boötes, on to Arcturus, to the right and left feet of Virgo, to the eighth degree of Libra; next it goes to the right hand of Centaurus, and, not far from the place where it touches the left hoof of Centaurus, it disappears from sight, to emerge again below Cetus; it then passes through his body and head and returns to the eighth degree of Aries.[29] (832)

The tropical colure originates at the eighth degree of Cancer, passes to the left front paw of Ursa Major, through his chest and neck, and crosses the celestial north pole; from here it goes through the hind parts of Ursa Minor, on through Draco and the left wing and neck of Cygnus, to touch the tip of Sagitta and the beak of Aquila; from this point it descends to the eighth degree of Capricorn and shortly plunges from view; it rises again below Argo, cuts through the rudder

[27] Martianus' tracing of the celestial equator generally conforms with that of Hyginus (*Astronomica* 4. 3) and almost exactly corresponds to that of Aratus *Phaenomena* 511-24. Aratus, however, says that the circle "has no share in Aquila."

[28] The tracing here corresponds to that of Aratus 501-6 and Hyginus 4. 4.

[29] Manilius (*Astronomica* 1. 603-17) also traces the colure, but with little correspondence to Martianus' tracing.

and upright stern, and returns to the eighth degree of Cancer.[30] (833)

The two oblique circles have breadth. The zodiac is a belt 12° wide.[31] Longitudinally it is divided into twelve segments (signs), each 30° in extent. Of the planets the sun (*sol*) is the only (*solus*) body to be borne in a course along the middle line (ecliptic) of this belt.[32] Twelve very conspicuous constellations lie within the zodiac. The other oblique circle, the Milky Way, is apprehensible by the eye as well as by the reason. Its breadth in many places diminishes below the normal width, but this loss is compensated by the great expanse in the stretch between Cassiopeia and the sting of Scorpio.[33] The last circle, the horizon, cannot be traced through the constellations, because it always changes with the rotation of the celestial sphere.[34] (834-36)

Apparently full of pride at her accomplishment of drawing the celestial circles, Astronomy now undertakes to fix the location of the five celestial parallels. Her bombast makes the matter—simple as it is when treated by other handbook authors—almost unintelligible:

> Now it is appropriate to explain what interval of distance or space has been admitted between the celestial circles by nature's intervention (*quid interstitii vel spatii intercapedo naturalis immiserit*). Between the arctic circle, which I have cut back in eight spaces, and the summer tropic, there is as much difference in space as between 8 and 6. Similar interjacent areas are contained in similar spaces (*idem interiectus spatiis similibus continetur*); thus it follows that one belt is larger than the other by one and a third times. Another intervening distance, between the summer tropic and the equator, is smaller than the belt above it as is the ratio of 4 to 6. The intervals of circles are reversed in the southern hemisphere [837].

What she is trying to say is that if a meridian circle is cut into 72 intervals—or a half circle, from north pole to south pole, into 36—the arctic circle is located 8 intervals from the pole, the Tropic of Cancer

[30] Again there is little correspondence to the tracing of Manilius 1. 618-30.

[31] The accepted modern figure is 16°.

[32] In placing *sol* and *solus* together, Martianus is assuming an etymological connection. Varro (*De lingua latina* 5. 68) and Cicero (*De natura deorum* 2. 68) derive *sol* from *solus*; and for other classical parallels see the note in the Pease edition of Cicero's work.

[33] Geminus (5. 69) says that it was because of the variations in breadth that many astronomers do not include this circle on their celestial globes.

[34] Geminus (5. 63), in remarking upon this difficulty, observes that the horizon can be represented by the stand which supports a globe.

6 intervals below that, and the equator 4 intervals below the tropic, with corresponding intervals marking the location of the corresponding parallels in the southern hemisphere.[35]

It is now time for Astronomy to introduce a catalogue of the constellations. These familiar objects, in addition to marking the location of the celestial circles, serve as reference points for observing planetary motions. According to custom, Martianus divides the constellations into the zodiacal constellations and those lying to the north or south of the zodiac. He does not see fit to list the signs of the zodiac—these are too well known to need enumeration—but he does say that although there are twelve equal zodiacal divisions, or signs, there are only eleven zodiacal constellations. Scorpio occupies its own space with its body and the space of Libra with its claws. The sign that the Romans call "Libra" (the Balance) Greek writers refer to as "the Claws". (839) (Astronomy's last statement is largely but not altogether correct.[36])

Martianus counts 19 constellations north of the zodiac: Ursa Major, Ursa Minor, Draco, Boötes, Corona Ariadnes, Hercules, Lyra, Cygnus, Cepheus, Cassiopeia, Perseus, Triangulum, Auriga, Andromeda, Pegasus, Serpentarius, Delphinus, Aquila, and Sagitta. These are the con-

[35] A more widely adopted scheme, the one used by Eratosthenes, assigned 60 intervals to the meridian, or 30 to a half circle. The latter was divided as follows: 6 intervals from the pole to the arctic circle; 5 intervals to the summer tropic; 4 intervals to the equator; and corresponding intervals for the circles in the southern hemisphere. Cf. Geminus 5. 45-46; Theon (ed. Hiller), pp. 202-3; Strabo 2. 5. 7; Achilles Tatius *Isagoge in Aratum* 1. 26; Manilius 1. 566-602; Macrobius *Commentary* 2. 6. 2-6.

[36] The Greek writers show a marked preference for the name *chelai* [the Claws]. Aratus uses this name throughout, as does Hipparchus, with the exception of one reference (*In Arati et Eudoxi Phaenomena commentarii* 1. 3. 5) to *zygos* [the Balance]. Ptolemy uses *chelai*, but is also found using *zygos* (*Tetrabiblos* 4. 4). Geminus (1. 1) uses *zygos*. Servius, in a comment typical of a "learned" Latin compiler, purports to trace (*ad Georgica* 1. 33) the discrepancy in systems to the original authorities. He attributed to the Egyptians the system of twelve signs, to the Chaldeans that of eleven. According to Servius the Chaldeans took Scorpio and Libra to be one sign, not seeking equality of extent for all the signs but having regard for the individual ranges of the signs, varying from 20° to 40° in extent, while the Egyptians preferred to consider all as being 30° in extent.

stellations recognized by Aratus, the classic authority in this field. Hyginus' list (*Astronomica* 2. 1) is similar to Martianus' and the order is almost the same.[37] Martianus disapproves of the practice of counting asterisms, those groups of stars resembling small animals or objects held or supported by the figures constituting the major constellations. Such star clusters as Capra (the Goat), which rests upon Auriga (the Charioteer), and Haedi (the Kids), which he holds in his arm, or Serpens, which Serpentarius grasps, or Panthera, which Centaurus carries, ought to be considered as parts of the more prominent constellations.[38] However, consistency is not a virtue of Martianus: having stated earlier that 35 constellations lie north or south of the zodiac, he now lists 14 southern constellations, giving a total of 33. His southern constellations are Hydra, Crater, Corvus, Procyon, Orion, Canis Major, Lepus, Eridanus, Cetus, Centaurus, Argo, Piscis Australis, Caelulum, and Ara. He adds that Aqua, which flows from the cup of Aquarius, and Canopus, also called Ptolemaeus, are more appropriately considered as parts of the constellations Aquarius and Eridanus. But these two, one of them a single star, must be counted as constellations to give a correct total.[39] (838)

Martianus' expressed intent of excluding asterisms from his list of

[37] The texts are compared in *Commentariorum in Aratum reliquiae*, ed. E. Maass, pp. xxviii-xxix.

[38] Uniformity in the ancient lists of constellations was not to be expected. The catalogues of some writers, like Vitruvius (*De architectura* 9. 4-5), are so confusing that the reader requires a chart, such as originally accompanied the catalogue, to understand it. Geminus (3.8) counts 22 northern constellations, adding to the traditional list Serpens held in the hand of Serpentarius, Coma Berenices, and Equuleus, the last on the authority of Hipparchus. Ptolemy (*Almagest* 7. 5), also following Hipparchus, counts 21, omitting Coma Berenices as an "unformed" (ἀμόρφωτος) asterism. Geminus (3.12) includes the Capra and Haedi supported by Auriga among the conspicuous weather signs which he adds to his list of constellations. The article "Constellation" in the eleventh edition of the *Encyclopaedia Britannica* presents a detailed historical account of catalogues of the constellations.

[39] Aratus lists 12 southern constellations, omitting Caelulum and Procyon on Martianus' list. Ptolemy lists 15, omitting Caelulum and adding Lupus and Corona Australis, not listed by Martianus. Geminus has 18, including 4 not found in Martianus' list: Aqua, Corona Australis, Lupus, and "Thyrsus-Lance" (*thyrsolongchos*) in the hand of Centaurus.

constellations is in keeping with his original avowal to be brief in handling the disciplines. Thus he also chooses to omit the next topic of writers on astronomy—the assignment of the constellations or parts of constellations to the five zones of the sky—because he considers the subject too complicated. Furthermore, he observes, the limbs of several of the mythological figures in the sky are mutilated by the celestial circles. The left hand of Boötes, for example, is assigned to the arctic zone, while the rest of his body is in the north temperate zone. Such details are disagreeable and depressing and are better left in obscurity. (840)

His feigned squeamishness about mutilated figures disappears in the next section as he takes up the subject of which constellations or parts of constellations are rising or setting as each of the twelve signs of the zodiac is rising. This precise information came down to Martianus' time in a Greco-Roman tradition, with little change, from Aratus' *Phaenomena* (*c.* 275 B.C.), a popular poem on the heavens which was designed to help readers understand astronomical allusions in poetry.[40] Although it contained no mathematics and avoided technicalities, the poem gained for its author a reputation as a great authority, even in Greece;[41] and in the Latin world Aratus was as well known to pupils in astronomy as Euclid was in geometry.[42] Martianus retains nearly

[40] The correspondences between passages in Aratus (569-732) and Martianus (841-43), even in matters of detail, are quite close.

[41] The only surviving work of Hipparchus, the greatest of the Greek astronomers, is his commentary on the *Phaenomena*. Had it not been for Aratus' reputation, we would have no work of Hipparchus at all. St. Paul's quotation from Aratus' invocation in Acts 17:28 is further evidence of Aratus' repute.

[42] Four Latin versions of the *Phaenomena* survive in part: that of Cicero (670 lines), of Germanicus Caesar (857 lines), of Avienus (1878 lines of paraphrase), and of Varro of Atax (only scant fragments preserved by Servius in his commentary on Vergil's *Georgics*). There are also four extant commentaries in Greek, by Hipparchus, Geminus, Achilles, and Leontius. The names of twenty-seven commentators are known. Aratus had a strong influence upon Lucretius and Vergil, and Ovid (*Amores* I. 15. 16) predicted that his fame would last as long as the sun and the moon. On the popularity of Aratus in Greek and Roman schools see Marrou, *History of Education*, pp. 185, 282, 434. On the methods of teaching astronomy, on the authors read in the schools, and on the prominence of Aratus as an authority see Hans Weinhold, *Die Astronomie in der antiken Schule.*

all the observational data of Aratus. Those pertaining to the rising of one sign will suffice as an example of the data for all twelve:

When Libra is rising, the remaining portions of Pegasus and Cygnus, the head of Andromeda, the shoulders of Cepheus, Cetus, and the meanders in the river Eridanus are setting. At the same time half of Corona, the right foot of Hercules, Boötes, all of Hydra except the end of the tail, and the equine part of Centaurus are rising [842].

The last set of observations, relating to Gemini's rising, was rejected by Denys Petau as an inept gloss; and Dick omits it from the text of his edition,[43] although it appears in full in Aratus' text.

The section that follows (844-45) contains some very interesting astronomical data, a scheme of rising times that is quite rare in ancient astronomical literature. Surprisingly for Martianus, the rising times are correct for the latitude implied by a statement he makes in the following section (846): that the longest day has 14 hours, 10 minutes, of daylight; and the shortest night, 9 hours, 50 minutes, of darkness. These observations hold true for a latitude north of Alexandria (14 hours) and south of Rhodes (14 1/2 hours); they do not refer to Babylon, whose longest day was always reckoned as 14 hours, 24 minutes.[44] Martianus then explains why different amounts of time are required for the rising and setting of zodiacal signs.

[43] See p. 444, apparatus. Cf. Aratus 724-31.

[44] I am indebted to Otto Neugebauer for pointing this out to me. At first Professor Neugebauer thought that Martianus might have computed the rising times himself or that he relied upon the observations of some North African astronomer. But the role of an observer or investigator in the interests of accuracy would be unthinkable for Martianus. He has no instincts for accuracy or consistency, and he must have gotten the data from some handbook tradition. Elsewhere he gives discrepant figures (13 and 12 2/3) for the hours of daylight at Meroë at the summer solstice, and he gives the same number of hours of daylight for Syene and Alexandria, though there is actually a difference of half an hour. Professor Neugebauer now thinks that no direct observation was involved here and that the originator of this scheme obtained the rising times by arithmetical interpolation. He characterizes the scheme as a botched-up System B and informs me that the only other example of this scheme of rising times known to him is found in *Catalogus Codicum Astrologorum Graecorum*, Vol. XII (Brussels, 1936), pp. 223-29. On System B see Neugebauer, pp. 158-60, 183-84. Leonardi "I codici" [1959], p. 482) points out that this passage (844-45) was one of two from Book VIII that were excerpted in fifteenth-century codices.

> Constellations that rise transversely and set vertically have swifter risings than settings; conversely, those that rise vertically and set transversely have slower risings than settings.[45] Cancer rises vertically and sets at an inclination, even though it has only a slight curvature in Capricorn. Cancer rises in $2^{1}/_{12}$ hours and sets in $1^{11}/_{12}$ hours. Here the difference is minimal. Leo rises in $2^{1}/_{3}$ hours and sets in $1^{2}/_{3}$ hours. Virgo rises in $2^{2}/_{3}$ hours and sets in $1^{1}/_{3}$ hours. The same holds for Libra. But Scorpio's rising time is less [than Virgo's] and the duration of its setting is greater; it rises in $2^{1}/_{3}$ hours and sets in $1^{2}/_{3}$ hours. Sagittarius rises in $2^{1}/_{12}$ hours and sets in $1^{11}/_{12}$ hours [844].

This is followed by the rising and setting times of the signs that rise transversely and set vertically. These anomalies he says, explain the inequalities in the duration of days and nights. As the sun moves into the signs that rise slowly, the days grow longer; when it enters the signs that rise quickly, the nights become longer.

Martianus notes that his readers may wonder how days and nights can vary in duration when all zodiacal signs extend over an equal amount of space and when, at all times, day and night, there are six signs above the horizon. There is no need for wonderment, he assures us. The observations are correct, and the equality of extent of the zodiacal signs is proved by measurements requiring the use of many clepsydras.[46] Although the signs consume varying amounts of time in their risings and settings, the sum is always the same if the rising and the setting of each of the signs are added together.[47] (846-47)

After these reassurances Martianus is ready to tackle another, more perplexing problem, and he speaks as if he were the first to find a solution.[48] Once again, if the signs occupy equal amounts of space and the sun moves at a uniform velocity, how is it that thirty-two days elapse during the sun's course in Gemini and twenty-eight days

[45] Cf. Geminus 7. 10-12; Cleomedes 1. 6. 31.

[46] Macrobius (*Commentary* 1. 21. 11-22) describes in detail the procedure of measuring the extent of the zodiacal signs by using clepsydras. In Geminus (1. 4) the division of the zodiac into twelve equal segments was demonstrated with a dioptra. Martianus later (860) mentions the use of clepsydras to measure planetary orbits.

[47] Geminus (7. 9-17) raises the same question and also points to the balancing of the pairs. He correctly attributes the difference to the obliquity of the ecliptic.

[48] This is characteristic of the Latin compilers. Cf. Macrobius' braggart tone when he pretends to refute Eratosthenes and Posidonius (*Commentary* 1. 20. 10) and Aristotle (*ibid.* 2. 15-16).

in Sagittarius, with a varying duration of days in the other signs? Here his explanation is better, though his boastful attitude is amusing, if not ludicrous. Up till now all men have supposed, the earth being at the center of the universe and the celestial sphere, that it is also at the center of the sun's orbit. This is obviously not true. The earth is not at the center of the sun's orbit, but is eccentric to it.[49] The obliquity of the ecliptic causes the sun, which moves along it, to be depressed or elevated in alternation, depending upon its juxtaposition with the signs. Cancer and Gemini are elevated in the "steeper"[50] regions of the sky, and Sagittarius and Capricorn are depressed, as they verge away. (848-49)

Martianus now comes to the planets and their motions; this subject is usually found last in the handbooks and occupies him for the remainder of Book VIII. The planets, he tells us, are seven in number. They require special attention because, whereas all heavenly bodies are swept along with the diurnal rotation of the celestial sphere, the planets have in addition their own motion in a reverse direction. The sun and the moon have been given countless names by the races of mankind. The other five planets are known by their mythological names and by the descriptive names given to them by the Greeks: Saturn, the "Shiner" (*Phaenon*); Jupiter, the "Blazer" (*Phaethon*); Mars, the "Fiery" one (*Pyroeis*); Venus, the "Light-bringer" (*Phosphoros*); and Mercury, the "Twinkler" (*Stilbon*).[51] These planets require varying amounts of time to make up the distance that they are carried backward in a single diurnal rotation of the sky: the moon a month; the sun a year; Saturn, the outermost, thirty years; and the intervening planets periods of time proportional to the distances they traverse in their orbits. (851-52) All seven planets are observed moving

[49] Geminus (1. 31-35) and Theon (ed. Hiller, pp. 152-57) give the same explanation for the anomaly.

[50] Macrobius (*Commentary* 1. 6. 51) and Cleomedes (2. 5. 113) speak of the "steeper" ascents of the sun in Gemini.

[51] The descriptive names, originated by the Greek astronomers, did not gain popularity, as Klibansky, Panofsky, and Saxl (p. 137) observe. The Romans did not attempt to translate these names. Cicero (*De natura deorum* 2. 52-53), like Martianus, uses the Greek forms; for other occurrences in classical literature see the notes in the Pease edition, II, 672-76.

toward the eastern horizon, yet their retrogressions are not directly counter to the direction of the diurnal rotation of the celestial sphere. This is fortunate for us, for the universe could not endure a contrary motion of its parts. (852-53)

It is the Peripatetics, according to Martianus, who refuse to believe in a countermotion of the planets; they believe that the planets are merely outdistanced, some greatly, some scarcely, by the swiftly moving celestial sphere. They are mistaken, however, and Martianus will refute them. It is not a question of swiftness or slowness or of motions over great or small distances. The motions of the planets must be considered individually; they will then be seen to be independent. They all differ in the times and circumstances of their periods. Five of the planets experience stations and retrogradations, but the sun and the moon are propelled in a steady course.[52] These two luminous bodies eclipse each other in turn; the other five are never eclipsed.[53] The three superior planets, along with the sun and the moon, have their orbits about the earth; Venus and Mercury, however, go about the sun. All seven planets make daily changes in their positions and orbits; no planet rises from the same position from which it rose on the previous day. The earth is eccentric to the orbits of all the planets. (853-55)

The sun, in moving from the summer tropic to the winter tropic, describes 183 circles across the sky; in moving back to the summer tropic it courses over the same circles. Mars, moreover, describes twice as many, Jupiter twelve times as many, and Saturn twenty-eight times as many, circles as the sun. These circles are referred to as parallels.[54] Here follows Martianus' unequivocal statement about the heliocentric

[52] Geminus (12. 19-22) attributes the lag theory to "many philosophers," and refutes it by pointing out that the fixed stars move in parallel courses and the planets, if they were merely being outdistanced, would stay in parallel courses. Instead the planets move obliquely across the zodiac, the moon going across its entire breadth. The stations and retrogradations of some planets are further proof of independent motion.

[53] It is surprising that little or no mention of the occultation of planets is found in the popular handbook literature. Aristotle (*De caelo* 2. 292a) reports an eclipse of Mars by the moon.

[54] Geminus (5. 12) states that there are 182 solar parallels between the tropics.

motions of Venus and Mercury, for which Copernicus expressed great admiration:[55]

Venus and Mercury, although they have daily risings and settings, do not travel about the earth at all; rather they encircle the sun in wider revolutions. The center of their orbits is set in the sun. As a result they are sometimes above the sun; more often they are beneath it, in a closer approximation to the earth.[56] Venus' greatest elongation from the sun is one and a half signs.[57] When both planets have a position above the sun, Mercury is closer to the earth; when they are below the sun, Venus is closer, inasmuch as it has a broader and more sweeping orbit [857].[58]

It is curious that the ancient writers who report Heraclides' theory of the heliocentric orbits of Venus and Mercury generally assign a fixed order to the planets, although a fixed order is irreconcilable with geoheliocentric motions. Sometimes they deal with both matters in succession, and occasionally they remark about the problems involved in reconciling the two theories. Calcidius attributes the order moon, Mercury, Venus, sun, to the Pythagoreans; and the order moon, sun, Mercury, Venus, to Eratosthenes. Later he expounds the Heraclidean theory, ascribing it to Heraclides by name.[59] Vitruvius gives the order as moon, Mercury, Venus, sun, and immediately thereafter vaguely describes the motions of Venus and Mercury as "wreathed about the rays of the sun, their center, as it were, themselves making stations and retrogradations."[60] Macrobius discourses learnedly and at length

[55] See n. 13 above.

[56] Mercury and Venus are half of the time above, half of the time below, the sun. On the earth side of the sun, these planets are more conspicuous, and this appears to be the cause of Martianus' confusion. Macrobius (*Commentary* 1. 19. 7) remarks on this very phenomenon.

[57] Or 45°. Later (882) he says 46°. The text is corrupt here, and there may have been a lacuna or a scribal error. It is unlikely that Martianus would have been guilty of so gross an error as to assign a maximum elongation of 45° to Mercury as well as to Venus, as Dick's text (451. 1-2) reads. In section 880 Martianus gives a maximum elongation of 22° to Mercury.

[58] On Martianus' and other reporters' statements about this theory, which was attributed in antiquity to Heraclides of Pontus, see Heath, *Aristarchus*, pp. 255-64. Pierre Duhem (III, 44-162) stresses the importance of these reports in keeping heliocentric views alive in the Middle Ages and traces the course of the theory in the medieval period.

[59] Calcidius 72-73, 110.

[60] *De architectura* 9. 1. 5-6.

about the two orders. He calls the one placing the sun second the "Egyptian" order, the other he calls the "Chaldean" order; and he explains the reason for the division of opinion. He then goes on to give an even vaguer statement than Vitruvius about the circles of Venus and Mercury about the sun.[61] Theon of Smyrna discusses rival views which place Mercury and Venus above or below the sun, changing the positions of Mercury and Venus with respect to each other, and later presents two epicyclic systems to explain the motions of Mercury and Venus, the second being like that attributed to Heraclides by Calcidius.[62] Even Ptolemy deals with the rival theories regarding their positions and the causes of the confusion, at the opening of Book IX of his *Almagest*. Ptolemy preferred the older view that placed them both beneath the sun at all times. Modern historians of astronomy have expressed surprise that Ptolemy did not see that the rival theories could be reconciled by abandoning his separate epicycles for Venus and Mercury and by making the sun the center of both their orbits, as Heraclides did.[63]

We should not be surprised, then, to find that, immediately after Martianus has described the heliocentric orbits of Venus and Mercury as alternately above and below the sun, he deals with the rival views of authorities who maintained a fixed order of the planets. Martianus, unlike the others, does not indicate a preference. (858)

Astronomy now tells the wedding guests that she is going to calculate the size of the orbits of all the planets, "an undertaking that astronomers consider a difficult one."[64] Once again Martianus reminds us of the bridesmaid's presence when he is about to depart from the standard handbook topics. Astronomy begins with the premise that the earth's circumference is 406,010 stadia. This figure, she says, was

[61] *Commentary* 1. 19. 1-6; tr. Stahl, pp. 162-64.

[62] Theon, ed. Hiller, pp. 140-43, 186-87.

[63] See J. B. J. Delambre, *Histoire de l'astronomie ancienne*, II (Paris, 1817), 265; and Dreyer, *History of Astronomy*, p. 201.

[64] Only Ptolemy attempted these calculations, but Martianus would not have known that. See Bernard R. Goldstein, "The Arabic Version of Ptolemy's *Planetary Hypotheses*," *Transactions of the American Philosophical Society*, Vol. LVII, pt. 4 (1967), pp. 5-12. Macrobius (*Commentary* 1. 20. 9-21), like Martianus, departs from handbook materials to explain the amazing procedures used by "the Egyptians" to ascertain the dimensions of the sun's orbit and orb.

offered by her sister Geometry and has been approved by Eratosthenes and Archimedes. We recall, however, that earlier Geometry correctly reported Eratotosthenes' figure for the earth's circumference as 252,000 stadia. Astronomy's startling figure alerts us to be wary of the disclosures to follow.[65]

She continues: Infallible reckonings show that the moon's orbit is one hundred times greater than the earth's circumference. The orbit is also found to be six hundred times as large as the moon itself. The true dimensions are ascertained by comparing the size of the moon with the breadth of the shadow cast on the earth when the moon lies directly beneath the sun during an eclipse.[66] If the reader is not becoming bored, Astronomy will explain how she obtained these figures.

When an eclipse of the sun occurs over Meroë, the entire orb is darkened there; but in the latitude of Rhodes, not far removed, the obscuration is only partial; and in the latitude of the mouth of the Dnieper the sun is not obstructed at all.[67] Since the distance from Meroë to Rhodes has been correctly ascertained, in stadia, Astronomy has calculated that the breadth of the shadow cast by the moon is one-eighteenth of the earth's circumference. (It should be pointed out in passing that I am making Astronomy's account plainer for modern readers than Martianus made it for medieval readers. His intended meaning would have been intelligible to those few readers who were familiar with the procedures described from having read about them in other authors.) Astronomy adds that bodies casting conical shadows are broader than their shadows and that thus she has determined from the distance between the latitudes at which the sun was partially obscured that the moon is three times as large as its shadow. Conse-

[65] The two different figures for the earth's circumference are perhaps the most astonishing of the many discrepancies in Martianus' work. It is not known who originated the larger figure. Aristotle (*De caelo* 2. 298b) reports an estimate of 400,000 stadia, current in his day; Archimedes (*Arenarius* 1; *Opera*, ed. J. L. Heiberg [Leipzig, 1913], II, 221) suggests 300,000 stadia as a reasonable estimate.

[66] In this would-be virtuoso demonstration, Astronomy is unaware that she is in effect comparing the moon's diameter with the earth's circumference—the disk of the moon with an arc of the earth's circumference.

[67] Cleomedes (2. 3. 95) has the sun totally eclipsed at the Hellespont and partially visible at Alexandria.

quently, she concludes, the moon is one-sixth as broad as the earth. (859)

The angular diameter of the moon is determined by comparing the amount of water that runs through a clepsydra during one complete rotation of the celestial sphere with the amount of water that runs through while the celestial arc occupied by the orb of the moon rises above the horizon.[68] The moon's orb is found to occupy $^1/_{600}$[69] of the complete circuit of the heavens. If the moon's angular diameter is $^1/_{600}$ of the circumference of the heavens, and the earth is 6 times as large as the moon, the moon's orbit will be 100 times as large as the earth. From this point it is a simple matter to measure the orbits of the other planets. Assuming that the planets travel at a uniform speed, the sun's orbit is 12 times as large as the moon's. Mars' orbit is then 24 times, Jupiter's 144 times, and Saturn's 336 times, as large as the moon's orbit. Taking the moon's orbit to be 100 times greater than the earth's circumference, Saturn's orbit is then 33,600 times greater than the circumference of the earth. (860-61)

Martianus, returning to conventional handbook topics, now takes up the orbits and behavior of each of the planets separately and in order, beginning with the moon. The cycle of lunar phases lasts a month, but the moon is always fully illuminated on the side facing the sun. On the thirtieth day of its cycle it reveals none of its light to us: it is between us and the sun, and the illuminated half faces the sun. The reason for the changing phases is that the moon's course is to one

[68] There are two versions of the procedure in the manuscripts. Petau rejected the longer version, which constitutes the bulk of section 860 in several editions, as a gloss drawn in substance from Macrobius *Commentary* 1. 21. 12-21. Actually the gloss is a verbatim quotation from Remigius' commentary on this passage (ed. Lutz, II, 277). Cleomedes (2. 1. 75) gives a brief account of this procedure as used to measure the apparent diameter of the sun; he attributes the method to the Egyptians, as does Macrobius.

[69] This amounts to an angular diameter of 36′, somewhat greater than either the actual figure or other ancient estimates:

Actual mean apparent diameter	31′ 59″
Cleomedes' figure	28′ 48″
Aristarchus' figure	30′
Ptolemy's mean figure	33′ 20″

Heath (*Aristarchus*, p. 314), citing Tannery's opinion, thinks that Martianus derived his figure from Varro and that it originated in a period before Hipparchus.

side of us and we glimpse the illuminated portion to an increasing degree as the moon moves away from the sun. At a position opposite the sun, the moon appears fully illuminated to us. Martianus gives the Greek terms for the phases: first appearance, *mēnoeidēs* [crescent-shaped]; at 90° eastward elongation, *dichotomos* [half]; at 135°, *amphikurtos* [gibbous]; and at 180°, *panselēnos* [full moon]. The same names are applied in inverse order as the moon diminishes in size on its return course.[70] In 24 hours the moon courses through 13° of its orbit. During the same interval the other planets course through the following portions of their orbits: Mars $1/2$°; Jupiter $1/12$°; and Saturn $1/28$°. (862-64)

The moon completes a circuit of the zodiac in $27\,2/3$ days[71] but requires $29\,1/2$ days[72] to overtake the sun, the reason being that while the moon is completing its orbit the sun has moved into the next sign, and sometimes into another sign beyond. For example, if the moon begins a cycle in the last degree of Libra, Scorpio, or Sagittarius, it does not overtake the sun again in the sign immediately following, but in the one after that. But because the sun tarries thirty or more days in diametrically opposite signs, and the moon overtakes the sun in $29\,1/2$ days, the moon will sometimes have conjunction with the sun twice in the same sign. The moon reaches the full phase on the fourteenth, the fifteenth, or, more frequently, the sixteenth day[73] of its cycle; but when a greater number of days elapse in the waxing, there will be fewer in the waning, so that the sum of days in a cycle is always the same. The period of a lunar year is 354 days (twelve conjunctions with the sun); a solar year exceeds a lunar year by 11 days, the difference being made up by intercalations. (865-66)

There are 12° of latitude in the belt of the zodiac, as pointed out above. Two planets have deviations through all 12°, one through as little as 3°, and the sun deviates from the ecliptic only in Libra, where it is deflected $1/2$° to the north of south.[74] One of the planets with

[70] Cf. Geminus 9. 11-12; Macrobius *Commentary* 1. 6. 55.

[71] A sidereal period.

[72] A synodic period

[73] Geminus (9. 14) gives the earliest day as the thirteenth, the latest as the seventeenth.

[74] Compilers of popular manuals continued to perpetuate this error, although

maximum deviation is the moon, which ranges 6° above or below the ecliptic. In its ascent and descent across the line of the ecliptic, it cuts the ecliptic at varying angles. The moon cannot return to the same position with respect to the sun—that is, to the same position in the same degree of latitude—until 235 months (nineteen years) have elapsed.[75] Fifty-five years are required for the moon to return to the same place on the same day of the year, in conjunction with the same fixed stars; and a lapse of a "great year" is required for all planets to return to their identical positions with respect to the fixed stars.[76] When the moon crosses the ecliptic in a northerly direction, it is said to be in ascending elevation; when returning to the ecliptic from the north, in descending elevation; when moving in a southerly direction from the ecliptic, in descending declination; and when returning to the ecliptic from the south, in ascending declination. (867-69)

These ascents and descents control the eclipses of the sun and the moon. If the ascending or descending moon crosses the ecliptic on the thirtieth day of its cycle, it lies directly beneath the sun with its entire body and causes an eclipse of the sun. This does not happen every month, because the moon is usually above or below the ecliptic on

---

Hipparchus had demonstrated that the sun does not veer from the ecliptic. The erroneous observation has been attributed to Eudoxus of Cnidus but may have originated earlier. On the error and its wide occurrence in antiquity see Heath, *Aristarchus*, pp. 198-200; Dreyer, *History of Astronomy*, pp. 94-95. To use this datum as evidence that a writer's astronomy was pre-Hipparchan is unwarranted. Compilers incorporated data of early and late vintage, as they found it, usually disregarding any inconsistencies that might be involved.

[75] Although Martianus has just referred to the moon's intersections of the ecliptic—known to astronomers as the ascending and descending lunar nodes—and although one complete revolution of the nodes in their westward movement along the ecliptic occurs every 18.6 years (approximately 18½ years according to Eudoxus, 18⅔ years according to Hipparchus), it appears that Martianus' authority is here referring to the "Metonic cycle" of 235 lunations which was proposed in the fifth century B.C. to bring solar and lunar years into agreement. On the Metonic cycle see Heath, *Aristarchus*, pp. 293-95.

[76] Martianus gives no figure for the duration of a great year, the estimates of which vary greatly with different writers. On classical references to the *magnus annus* see the Pease edition of Cicero's *De natura deorum*, II, 668-69. For survivals of the concept see "Magnus annus" in the indexes to the volumes of Thorndike's *History of Magic and Experimental Science*.

the thirtieth day. In like manner the moon is eclipsed when it crosses the ecliptic on the fifteenth day, in a position of opposition to the sun. The sun projects its shadow along the ecliptic and, if the moon reaches this line on the fifteenth day, a lunar eclipse occurs.[77] Eclipses cannot recur within six months,[78] for the moon will not be found on the ecliptic twice on the fifteenth or first days of its cycles during that period:

If, in returning to the ecliptic from the north, it comes into close lateral proximity with the sun but does not move into an obstructing position, it is said to produce an approximation in transit (παράλλαξιν ἐν συνόδῳ); but if, in coming from the north, it does move into conjunction and obstructs the sun, it is said to produce an eclipse in northern transit (καταβίβασιν ἐν βορείῳ συνόδῳ). If it comes from the south and does not move into conjunction, it produces an approximation in southern transit (παράλλάξιν ἐν συνόδῳ νοτίῳ); and if, in returning to the ecliptic from the south, it crosses the path of the sun, it produces an ascending eclipse node (ἀναβιβάζοντα σύνδετμον).[79]

Martianus concludes his account of the moon's motions by remarking that the vagaries of this planet confound mortals with their complexity, a statement which still holds true. (869-71)

The sun, like the moon, has a twofold motion, being swept along in an east-west direction by the diurnal rotation of the celestial sphere and maintaining its proper motion obliquely along the ecliptic in an easterly direction. Its daily shifts in points of rising cause it to describe 183 circles as it ranges obliquely from the summer tropic to the winter tropic. In its return course from the winter tropic it describes the very same circles. Each circle cuts across the zodiac twice and is drawn through signs opposite each other: the first circle of Aries is also the first of Libra; the thirtieth of Aries is the thirtieth of Libra; the first of Taurus is the first of Scorpio, and so on. The sun traverses these circles, in both ascent and descent, in 365$^1/_4$ days. It is interesting

[77] Cf. Geminus chaps. 10, 11; Cleomedes 2. 6. 115-21; Theon (ed. Hiller), pp. 193-97.

[78] A. Pannekoek, *A History of Astronomy* (London, 1961), p. 46, gives the reason and points out that the ancient Babylonians were aware of this fact.

[79] Cf. Ammianus Marcellinus *Res gestae* 20. 3. 4; Ptolemy *Almagest* 6. 6; Favonius Eulogius *De somnio Scipionis* 9 (ed. van Weddingen [Brussels, 1957] 21. 20 - 23. 4)—Van Weddingen mistakenly interprets the passage in Favonius as referring to epicycles and eccentrics.

that, although the northern and southern hemispheres are of the same dimensions, and the signs extend equally on either side of the equator, the sun courses through the signs in unequal periods. It completes its ascending course to the summer tropic in 185 1/4 days and its descending course in 180 days.[80] The reason for this anomaly is that the earth is eccentric to the sun's orbit, which is more remote (from the earth) in the upper hemisphere. The sun, as it moves upward toward Cancer, gradually brings on warmth; we have the scorching heat of summer while it is in Cancer; and as it moves southward toward Capricorn, days become chillier. For antipodeans the seasons are reversed: summer when the sun is in Capricorn, winter when it is in Cancer. When the sun is at the equator, both temperate zones have mild weather. (872-74)

The discussion of the sun's orbit brings Martianus to the subject of the hours of daylight at the different "climates" or latitudes. Winter nights correspond exactly in length to summer days, and summer nights to winter days. On two days in the year the hours of daylight and darkness are exactly equal. The longest day has 14 equinoctial hours, the shortest 9,[81] but the hours vary according to latitude. There are eight climates, beginning with the one passing through Meroë, which is closest to the summer tropic.[82] North of it we come in sequence upon the climates passing through Syene, Alexandria, Rhodes

[80] Geminus (I. 13-17, 31-35) says 184 5/8 days elapse in the ascending course and 180 5/8 in the descending; Theon (ed. Hiller), p. 153, has 187 and 178 1/4 days.

[81] Most manuscripts read *novem*, and this is the reading adopted by Eyssenhardt. Apparently some scribes were disturbed because 14 and 9 do not give a total of 24 and because Martianus' statement here is inconsistent with his remark (877) that the longest day at the climate below the Rhipaean Mountains has 16 hours of daylight. Thus Dick adopts the emended reading of some manuscripts: 16 hours of daylight at the summer solstice and 8 hours at the winter solstice. But Martianus is not concerned about fractions of hours, so we should not be concerned about a total of 23. It crops up again (877) for the Hellespont climate. Martianus has in mind not the longest recorded day at the northernmost climate but the longest day at the latitude he was referring to in section 846, when he said that the longest day has 14 (and a fraction) hours, and the shortest day 9 (and a fraction) hours. On the corruptions that commonly appear in manuscripts in the handling of Roman numerals see above, p. 141.

[82] Martianus is confused here. Classical writers since Eratosthenes had placed Syene at the tropic.

(the fourth and middle one[83]), Rome and Macedonia, the Hellespont and Gaul, Germany and the mouth of the Dnieper, and finally through the region above the Sea of Azov and below the Rhipaean Mountains.[84] (875-76)

The maximum and minimum hours of daylight given by Martianus (877) for each of the eight latitudes are summarized here in tabular form.[85]

Maximum and Minimum Hours of Daylight

Meroë	13	11
Syene*	14	10
Alexandria	14	10
Rhodes*	14	9
Rome	15	9
Hellespont*	15	8
Dnieper	16	8
Rhipaean Mts.	16	8

* Hipparchus' figures for Syene, Rhodes, and the Hellespont (i.e., Pontus) were given to the half hour, according to Strabo: Syene, 13½; Rhodes, 14½; the Hellespont, 15½ (Strabo 2. 5. 36, 39, and 41, respectively). Honigmann believes that *Diahellespontu* is a corruption of *dia mesou Pontou.*

[83] A middle climate among eight is of course impossible. In the classical scheme of Eratosthenes and Hipparchus, Rhodes marked the middle and principal latitude of the seven climates. Martianus added an eighth climate, below the Rhipaean Mountains; hence his confusion.

[84] We have to depend mainly upon Strabo (2. 5. 35-42) for information about the establishment of the principal climates. Strabo's report is not clear or consistent because of his penchant for criticizing his authorities. He informs us that Hipparchus adopted the principal climates of Eratosthenes and recorded the hours of daylight for each, which vary by half hours in duration. E. Honigmann's *Die sieben Klimata* is a careful study of the ancient references to the climates. See also Bunbury, II, 4-14.

[85] Honigmann (pp. 50-52), believes that Martianus' scheme of climates stems from Varro, for, unlike Pliny, Martianus uses the Greek word for climates and gives Greek designations to them, even to the one through Rome (διὰ Ῥώμη). He notes that the old Eratosthenean Klima VI, through the middle of Pontus, has been removed, and Klima V, through the Hellespont, then becomes VI, making room for the climate through Rome to become V. This leads him to conclude

As one draws near the pole, days become longer and nights shorter; consequently it is to be assumed that there is perpetual daylight at the pole.[86] Increments in the amount of daylight occur as follows: $^1/_{12}$ of the total increase from the winter solstice to the summer solstice is added in the first month, $^1/_6$ in the second, $^1/_4$ in the third, $^1/_4$ in the fourth, $^1/_6$ in the fifth, and $^1/_{12}$ in the sixth.[87] The reason for the variation is that the zodiac winds around Cancer and Capricorn but cuts across the equator almost directly (878). So much for the sun and the moon.

Mercury and Venus have their orbits about the sun, "off to one side, in a certain manner," and do not encompass the earth in their orbits. They are impelled back and forth alternately. They are observed rising and setting because they are swept along by the motion of the celestial sphere.

Mercury requires nearly a year to complete its orbit and moves through 8° of latitude. Its maximum elongation is 22°;[88] never does it get as far away as two signs from the sun, ahead or behind it. Mercury therefore never has acronical risings, for these occur only to planets that are situated diametrically opposite the sun.[89] Mercury's risings are inconspicuous and brief:[90] one when elongation permits and the planet is not obliterated by the sun's brilliant rays; a second when, as it retro-

---

that the table of climates had been revised by some Western-oriented compiler, possibly Posidonius or Nigidius Figulus. See also Miller, VI, 141.

[86] Pliny also exhibits confusion about the duration of daylight in the far north. He correctly reports (4.104) 24 hours of daylight at Thule at the summer solstice, but he incorrectly reports (2. 186-87) Pytheas as stating that there are six months of daylight at Thule in the summer season. Geminus 6. 13-15) shows a correct understanding of the duration of daylight at the pole.

[87] This set of increments agrees exactly with that of Cleomedes (1. 6. 27).

[88] Pliny gives Mercury's maximum elongation first as 22° in Roman numerals (2. 39), then as 23° with the numbers written out (2. 73). Theon (ed. Hiller), p. 137, and Calcidius 70 give it as 20°. For other classical references on Mercury's elongation and on other aspects of the planet's behavior see the Pease edition of Cicero's *De natura deorum*, II, 674-76.

[89] Theon (ed. Hiller), p. 137, explains acronical risings in the same way.

[90] Habitual sky-watchers may pass a lifetime without being certain of having glimpsed Mercury. Copernicus is said never to have seen it. Observation of it is made easy when it is in proximity to Venus, as it was during the month of January, 1965.

grades in the west, it moves into the vicinity of the sun and fades from sight. These first and last visibilities recur in the fourth month, and not always then. (879-81)

Venus also completes its orbit in a period of about a year. Located on its own epicycle, it goes about the sun, which it sometimes passes, sometimes lags behind. When Venus is in retrograde motion, it takes longer than a year to traverse its orbit; but when it is going in direct motion, it completes its course in as little as eleven months.[91] When it rises ahead of the sun it is called Lucifer; when it blazes in the evening, after sundown, it is called Vesper. Venus, like the moon, deviates through all 12° of the zodiac's latitude. Its maximum elongation is 46°.[92] Venus is the only one of the five planets, like the moon, to cast a shadow[93] and the only one to linger for a long time before yielding to the sun's brilliance. In morning risings it frequently tarries for four months, but in the west never more than 20 days.[94] Its risings and last visibilities recur in nine- to ten-month cycles. (882-83)

Mars has its own orbit, beyond the sun's and about the earth, which is eccentric to that orbit. Mars completes a revolution in approximately two years. It has a motion in latitude of 5°. Like the two planets beyond it, it experiences stations and retrogradations, but it has its own apogee, first station, and exaltation apart from the others. Its apogee, the point where its orbit reaches highest elevation above the earth, is in Leo.[95] Its first station is unique. Because Mars' orbit is close to the

[91] Among the Greek popular handbook authors, Geminus, Cleomedes, Achilles Tatius, and pseudo-Aristotle give the periods of Mercury and Venus as a year or approximately a year, as does Cicero (*De natura deorum* 2. 53). See note in the Pease edition, II, 675. Vitruvius (9. 1. 8-9) gives 360 days for Mercury and 485 for Venus; Pliny (2. 38-39) gives 339 and 348 days. W. L. Lorimer (*Some Notes on the Text of ps. Aristotle "De Mundo"* [Oxford, 1925], pp. 129-30), has prepared a table of times given by ancient authors. Martianus gives the correct explanation of the variations in time.

[92] Pliny (2. 38) has 46°; Theon (ed. Hiller), p. 137, and Calcidius 70 have 50°. The correct figure is approximately 45°.

[93] Cf. Pliny 2. 37.

[94] An amazing statement for an ancient author to make, considering that celestial phenomena were much more familiar to the ancients than they are to us, and that the lingering brilliance of Venus in the western skies is one of the most conspicuous of all celestial phenomena. Maximum brilliance is attained about 36 days preceding and following inferior conjunction.

[95] Cf. Pliny 2. 64.

sun's, it feels the effects of the sun's rays even at a position of quadrature and comes to a halt at a distance of 90° from the sun on either side.[96] Mars' exaltation occurs in the twenty-ninth degree of Capricorn.[97] (884)

Propitious Jupiter, being higher than the aforementioned planets, completes its revolution in twelve years and has a deviation in latitude of 5°. Its apogee occurs in Virgo and its exaltation in the fifteenth degree of Cancer.[98] Its elevations and depressions prove that the earth is eccentric to its orbit. (885)

Saturn, the outermost of the planets, completes a revolution in slightly less than thirty years and deviates in latitude 3° or sometimes only 2°.[99] It has its apogee in Scorpio and its exaltation in the twentieth degree of Libra.[100] (885)

[96] Cf. Pliny 2. 60, and note the correspondence of texts: *etiam ex quadrato sentit radios* (Pliny); *etiam in quadratura . . . radios sentit* (Martianus). The numerous correspondences between Pliny and Martianus in these closing sections of Book VIII suggest that their data originated in the same astronomical manual, presumably that of Varro.

[97] Pliny (2. 65) says the twenty-eighth degree, according to the figure adopted by the editors of the Budé and Loeb editions. J. Beaujeu, the Budé editor, *Pline l'Ancienne, Histoire Naturelle Livre II*, p. 28, lists in his apparatus the reading XXVIIII (sic) in several manuscripts. The manuscripts of Pliny use Roman numerals, whereas those of Martianus write the figure out. Martianus' source was using an older manuscript of Pliny than any now extant. It would appear that Martianus' figure is the correct one for Pliny, another instance of the value of Martianus in emending Pliny's text. Mommsen pointed out that Solinus' correspondences with Pliny's geographical books would enable an editor to emend Pliny's text.

[98] These observations correspond to Pliny 2. 64-65.

[99] It will be of interest to compare here Martianus' figures for planetary deviations in latitude with those of Theon (ed. Hiller), p. 135; Cleomedes 2. 7. 125; and Pliny 2. 66-67:

	Martianus	*Theon*	*Cleomedes*	*Pliny*
Sun	1°	1°		2°
Moon	12°	12°	12°	12°
Venus	12°	12°	10°	14°
Mercury	8°	8°	8°	8°
Mars	5°	5°	5°	4°
Jupiter	5°	5°	5°	3°
Saturn	2-3°	3°	2°	2°

[100] Again the observations correspond to those of Pliny 2. 64-65.

Martianus concludes his book on astronomy by redefining the terms applied to the superior planets. Their "last visibilities" occur when their shimmering light finally disappears in the glare of the sun. These planets have their "morning (first) stations" at 120° distance from the sun and their "evening risings" at opposition (180°); likewise, on the other side, they have the "evening (second) stations" at a position 120° away from the sun. Within 12° the rays of the sun engulf and obliterate the light of the superior planets. It is the powerful effect of the sun's rays that is responsible for the anomalies in the orbits of all the planets—the stations, retrogradations, progressions, elevations, and depressions. As the rays strike the planets, they cause them to rise on high or to be depressed, to deviate in latitude or to retrograde.[101] (887)

Book VIII ends abruptly in a lacuna. It appears that nothing important has been lost from the content of the discipline, since Martianus, either here or in Book VI, has taken up all the conventional topics of an *ars astronomiae*. What is regrettable is the loss of the closing scene in the celestial hall. Astronomy makes the best presentation of the quadrivium bridesmaids, perhaps of all seven bridesmaids. It is safe to assume that Archimedes and Ptolemy would be duly impressed with her eloquent discourse and would join in the applause.

[101] Doctrines about the sun as the regulator of the planets and about the effects of its powerful rays were widely circulated by popular writers on cosmography. Pliny (2. 69-70) says that the planets, when struck by a triangular ray (i.e., at 120° elongation), are deflected in their courses and lifted straight upward, thus appearing to us on earth to be at station. Vitruvius (9. 1. 12) also points to the powerful effect of the sun's rays at trine aspect. Macrobius (*Commentary* 1. 20. 4-5) speaks of the sun's power and influence in determining the limits beyond which the planets cannot recede and are seen to turn back toward the sun; Lucan (*Pharsalia* 10. 201-3) says the same thing in verse. Theon (ed. Hiller), pp. 187-88, describes the sun as the animating principle and heart of the universe. William of Conches (*De philosophia mundi* 2. 23 [64d]), as is his wont, takes the popular classical doctrine and gives it his own explanation: a planet, because of the sun's rays, loses so much moisture that it rises in space, and when recondensation takes place it returns to its regular orbit. One wonders how much this classical notion, sometimes referred to as the "attraction-repulsion" theory, was influenced by the Stoic Cleanthes' view that the sun is the ruling force of the universe and how much by the heliocentric theories of Heraclides and Aristarchus. See R. M. Jones, "Posidonius and Solar Eschatology," *CP*, XXVII (1932), 113-35; Duhem I, 441-46; and Erhardt-Siebold and Erhardt, *Astronomy of Erigena*, pp. 25, 65.

On Harmony

AS THE STORY resumes Venus is once again bitterly complaining about the tedium of the bridesmaids' arduously discoursing on learned subjects at a wedding celebration that calls for song and wanton mirth. Gay choruses are her way of life; she cannot bear the dronings of somber maidens steeped in Attic lore. Her attention till now has been on diverting the males in the audience. Mars and she have been exchanging knowing glances, Bacchus is becoming interested, and momentarily the wedding hangs in the balance as Mercury is transported by her charms and is ready to jilt his bride Philology. Venus retains all her pagan attractions for the sensuous Martianus. Drastic action is called for to conclude the wedding and the liberal arts. At this point Apollo and Minerva intervene in the role of patrons of the arts.

A seventh bridesmaid is still to be heard from, Harmony, the darling of the heavenly gods.[1] Then there are Medicine and Architecture standing to one side and, as Apollo observes, eager to hold forth. Varro had sanctioned these studies by incorporating them in his *Nine Books of the Disciplines*. But Martianus thought otherwise, and ever afterward these subjects were excluded from the liberal canon and were regarded as professional disciplines. Apollo's suggestion is quickly rejected. The Varronian bridesmaids are concerned with mundane matters that should have no hearing before a celestial company. It is getting late and the wedding guests are becoming uneasy as the undaunted Apollo introduces seven more maidens, more impressive than

[1] Because the motions of the heavenly spheres produce harmony. E. A. Lippmann stresses the importance of arithmetic and music in medieval traditions of the quadrivium in "The Place of Music in the System of the Liberal Arts," in *Aspects of Medieval and Renaissance Music*, pp. 545-59. He aptly points out (p. 550) that for Boethius, Nicomachus, Iamblichus, and Augustine the only quadrivium treatises which have come down to us are on arithmetic and music. The discipline portion (920-96) of Book IX has been translated, with introduction and notes, by F. H. Copp as an M. A. thesis, Cornell University, 1937 (unpublished).

the first group in fact, since they are privy to secrets known only to divinities. These are the seven prophetic arts. Apollo is in favor of hearing them all. Never in classical epic was Jupiter called upon to arbitrate between disputants more widely divergent than Venus and Apollo. His response is quick and decisive. Harmony is to be presented at once and the wedding then to be concluded; the other maidens are to be held over for some future occasion.

Harmony's appearance calls for a rousing prothalamium. Venus, quickened by the sudden turn, joins the festivities. Harmony is her daughter and she calls upon Hymenaeus to sing the prothalamium. The introduction of Harmony and the approaching wedding ceremonial provide Martianus with his best opportunity to display his poetic talents. The prothalamium (902-3) and Harmony's response (915-19), with its recurring refrain recalling the *cras amet* of the *Pervigilium Veneris*, reflect no great discredit upon the classical models that inspired him.

Harmony herself is ineffably dazzling and Martianus is stricken in his efforts to describe her. A lofty figure, her head aglitter with gold ornaments, she walks along between Apollo and Athena. Her garment is stiff with incised and laminated gold; it tinkles softly and soothingly with every measured step. She carries in her right hand what appears to be a shield, circular in form. It contains many concentric circles, and the whole is embroidered with striking figures. The circular chords encompass one another and from them pours forth a concord of all the tones. Small models of theatrical instruments,[2] wrought of gold, hang suspended from Harmony's left hand. No known instrument produces sounds to compare with those coming from the strange rounded form. As Harmony enters the hall a concord swells from the shield. "Jupiter recognizes the exalted strains as honoring a certain secret fire and inextinguishable flame and the heavenly throng rise in reverence and homage to supramundane intelligence." We incline to agree with Martianus when he concludes his description of this mys-

[2] Remigius *ad loc.* (ed. Lutz, II, 313) rightly observes that the term refers to musical instruments made by mortals, as contrasted with the divine music symbolized by the shield. Needless to say, archaeology has recovered no examples of a shield with circular chords.

tical scene by referring to it as "indescribably stirring" (*egersimon ineffabile*). (909-10)

The cordial reception given to her songs encourages Harmony to submit to the test of learning undergone by her sisters and to present the precepts of her discipline. She is filled with trepidation. Ages ago she was detestable to mortals, and only now is the darkness beginning to be dispelled. The Creator begot her as one of the celestial twin sisters (the other twin being Arithmetic).[3] She followed the swirling motions of the spheres and assigned tones to the swiftly moving planets.

> But when the Monad and first hypostasis of intellectual light was conveying souls that emanated from their original source to earthly habitations, I was ordered to descend with them to be their governess. It was I who assigned the numerical ratios of perceptible motions and the impulses of perfect will, introducing restraint and harmony into all things [922].[4]

The conventional introduction for ancient manuals on music and for excursuses or chapters on music written by encyclopedists consisted of a brief discourse on the origins of music and a recital of examples demonstrating the universal power of music.[5] Harmony is here weaving Neoplatonic doctrines into her narrative to produce such an introduction. The quotation above is the most Neoplatonic passage in the entire *Marriage*, but it does not constitute a sufficient basis for regarding Martianus as a representative of the Neoplatonic

[3] Remigius (ed. Lutz, II, 322) offers this explanation: harmony is intrinsic to the heavens, and Harmony and Arithmetic are twin sisters because there are numerical ratios in harmony.

[4] *Sed cum illa monas intellectualisque lucis prima formatio animas fontibus emanantes in terrarum habitacula rigaret, moderatrix earum iussa sum demeare. Denique numeros cogitabilium motionum totiusque voluntatis impulsus ipso rerum dispensans congruentiam temperabam.*

[5] Examples of these stereotypes are very numerous. A few are of particular interest because they show the commonplace character of these materials more clearly; see Macrobius *Commentary* 2. 3. 1-10; Censorinus *De die natali* 13-14; Sextus Empiricus *Adversus mathematicos* 6. 7-18, 32; Cassiodorus 2. 5. 1-3, 9; Isidore 3. 15-17. The last two cite biblical instances together with the stock examples from classical sources. For other classical references and some recent studies on these stereotypes see Fontaine, I, 421-25.

school.[6] Martianus' purview did not extend beyond the elements of the disciplines; when he comes upon any complex or abstruse matters he usually remarks that they befit Attic sages and are to be avoided. Statements about the Neoplatonic hypostases and Monad became commonplaces in the writings of compilers who knew little else about Neoplatonism.[7]

Still somewhat apologetic, Harmony recites a long list of instances revealing the strange and mystical powers of music. She points out that many nations and peoples are stirred to combat or brandish weapons to the sounds of musical instruments—Cretans, Lacedemonians, Sybarites,[8] Amazons, to cite a few. One Amazon woman, who had approached Alexander in the hope of conceiving by him, was given a piper instead and went away elated with her gift. The trumpet (*tuba*) arouses the spirits of steeds of war and also sharpens the keen edge of wrestlers and other competitors in public games. The therapeutic effects of music[9] are common knowledge, such as the effect of incantations and instrumental music upon mental patients, trumpet blasts upon deaf patients, cithara playing upon rampant pestilence, and the soothing effect of the aulos on gout of the hip. Herophilus, the eminent Alexandrian physician, checked the condition of his patients by the rhythm of their pulse beat.[10]

Animals, too, are sensitive to music. Stags are caught through the use of shepherds' pipes, fish stop swimming at clattering sounds, swans are attracted to the cithara, other birds to pipes, Indian elephants

[6] See above, p. 10.

[7] The Monad and hypostases also figure prominently in the writings of arithmologists, such as Nicomachus, who antedated formal Neoplatonism. See *Nicomachus*, tr. D'Ooge, pp. 95-99.

[8] According to legend the Sybarites taught their cavalry horses to dance to the music of reed pipes. During a battle between Sybaris and neighboring Croton, the invading Crotonians allegedly played on the pipes and the Sybarite horses danced off with their riders.

[9] Martianus is thinking of music in its classical conception as embracing both vocal and instrumental music and rhythm and rhythmic recitation.

[10] Remigius *ad loc.* (ed. Lutz, II, 327) says that if the pulse beat is rhythmic, the patient is normal. The use of music in treating mental illness was first suggested by Theophrastus, according to Klibansky, Panofsky, and Saxl, p. 46. On Pythagoras' belief in the healing and cleansing effects of music see C. J. de Vogel, *Pythagoras and Early Pythagoreanism*, pp. 164-65.

respond to instrumental music, and cobras are charmed, their bodies known to burst asunder from the effect. Trees and crops are lured[11] by music, and the moon is eclipsed by incantation. Varro, a recent reporter,[12] claims to have observed islands in Lydia float away from the mainland at the first strains of tibias, circle around to the middle of the lake, and return to the mainland. Harmony could continue to recall her countless benefactions to mankind but she must get on with the subject matter of her discipline in order to fulfill her promise to the bride. (925-29)

Harmony opens her discipline with a series of classifications and definitions. First she explains that hers is the art of using good proportions in rhythm and melody[13] and that she intends to take up melody first. The treatment that follows, wanting as it may appear in comparison with the Greek treatise from which it was largely derived, had sufficient merit to rank its author as the second most important Latin authority on music, after Boethius.

Sounds that strike the ear in right proportion (*rite*) form either a whole tone (*tonus*), a half tone (*hemitonium*), or a quartertone (*diesis*).[14] A whole tone is an interval of appropriate size between two mutually different sounds. The interval that lies in the middle of a whole tone is called a half tone.[15] Among quarter tones there are three

[11] I.e., from the owner's property.

[12] Varro would have been a recent reporter to the first compiler who drew this bit of "information" from him.

[13] 930: *officium meum est bene modulandi sollertia, quae rhythmicis et melicis astructionibus continetur*. The definition of music as *scientia bene modulandi* is found in Augustine and Cassiodorus and may have been Varro's definition. See Fontaine, I, 419-20. Terms like *bene modulandi* and *modulatio* are hard to define. There has been so much confusion and debate among classical musicologists about the correct interpretation of ancient musical terminology that one is inclined to retain the Latin terms and not to attempt to translate them. I show the Latin in parentheses in the interest of clarity.

[14] On the controversial subject of the meaning of *tonos* in Greek musical theory and on recent interpretations see W. D. Anderson, pp. 30-31, 226-27. The systematization that included quarter tones was a theoretical one. See J. F. Mountford and R. P. Winnington-Ingram, "Music," *OCD*, p. 587.

[15] Because the ratio 9:8 (whole tone) has no rational square root, there cannot be a mathematically exact half tone. This was known to the early Pythagoreans. Cf. Euclid *Sectio canonis* 16; Theon, ed. Hiller, p. 70; Macrobius *Commentary*

sizes: the first, a smaller one, is called *tetartemoria* by reason of the fact that it measures a "fourth part" of a whole tone; it is also known as the "enharmonic interval" because it is a basic step in constructing the enharmonic genus[16] of musical movement (*modulandi*); the second, a larger interval, is called *tritemoria*, consisting of a "third part" of a tone; it is known as the "chromatic interval" because it is the basis of the chromatic genus;[17] the third size consists of a quarter part of a tone and a half part of a quarter tone, and is called the *hemiolia* of the enharmonic division.[18] (930)

"Note" or "key" (*tonus*) and "sound" (*sonus*) are generally interchangeable in their applications. In every transposition scale (*tropus*) there are eighteen *tonoi*. (See Table 1. Martianus gives both the Greek and Latin names. The Greek word *hypate* means "highest" and *nete* means "lowest," but the terms are taken from the position of the hand on the string rather than from the pitch of the notes produced.[19] *Hypate* is actually the lowest note and *nete* the highest note, and they are so translated here.) (931)

All musical movement (*modulatio*) consists of lower- or higher-pitched tones.[20] Low pitch refers to a tone that relaxes through a slackening of the sound; a high pitch is a tightening of the music to a thin and shrill sound. There are three consonances (*symphoniae*) in every octave species (*tropus*). (See Table 2.)

---

2. 1. 21. For a discussion of the mathematics involved see A. Wasserstein, "Theaetetus and the History of the Theory of Numbers," *Classical Quarterly*, new ser., VIII (1958), 173.

[16] The enharmonic tetrachord consists of a ditone and two quartertones: 2, $^1/_4$, $^1/_4$.

[17] The chromatic tetrachord, with these intervals: $1^5/_6$, $^1/_3$, $^1/_3$.

[18] Mountford and Winnington-Ingram, p. 587, place this interval in the hemiolic chromatic genus. The intervals are $1^3/_4$, $^3/_8$, $^3/_8$.

[19] Isobel Henderson, in the chapter "Ancient Greek Music" in *The New Oxford History of Music*, I, 345, explains the names as referring to the position of the hand in playing the instrument, *hypate* meaning the "highest" string to the hand on the tilted cithara.

[20] H. M. Klinkenberg, "Der Verfall des Quadriviums im frühen Mittelalter," in Joseph Koch, ed., *Artes Liberales von der antiken Bildung zur Wissenschaft des Mittelalters*, p. 8, points out that Martianus, in contrast to Augustine, defines *modulatio* only in the sense of audible music.

Table 1. System of Eighteen *Tonoi**

Greek Name	*Latin Name*	*Translation*
Nete hyperbolaion	Ultima excellentium	Highest tone of the highest tetrachord
Hyperbolaion diatonos†	Excellentium extenta	Extended tone of the highest tetrachord
Trite hyperbolaion	Tertia excellentium	Third tone of the highest tetrachord
Nete diezeugmenon	Ultima divisarum	Highest tone of the disjunct tetrachord
Diezeugmenon diatonos	Divisarum extenta	Extended tone of the disjunct tetrachord
Trite diezeugmenon	Tertia divisarum	Third tone of the disjunct tetrachord
Paramese	Prope media	Tone next to the middle tone
Nete synemmenon	Ultima coniunctarum	Highest tone of the conjunct tetrachord
Synemmenon diatonos	Coniunctarum extenta	Extended tone of the conjunct tetrachord
Trite synemmenon	Tertia coniunctarum	Third tone of the conjunct tetrachord
Mese	Media	Middle tone
Meson diatonos	Mediarum extenta	Extended tone of the tetrachord of the middle
Parhypate meson	Subprincipalis mediarum	Next to the highest of the tetrachord of the middle

* In this table (compiled from Martianus 931) the Greater and Lesser Perfect Systems are combined, producing the Immutable System. To the Greater Perfect System is added the tetrachord synemmenon from the Lesser Perfect System. See Mountford and Winnington-Ingram, pp. 585-86. Vitruvius (5. 4. 5) lists the same eighteen *tonoi* given here. (Vitruvius also recommends a knowledge of music to military engineers as useful in determining the proper tension of the sinews of catapults, by their pitch when vibrated, and to theater architects in the location and design of theaters. His expedient of placing bronze acoustical vases in theaters to resonate the voices of actors was tried in the choir lofts of many medieval churches. See George Sarton, *Introduction to the History of Science*, III, 1569-70. Resonators are still being built into modern music halls.)

† Most ancient writers prefer the term *lichanos* [forefinger string] to *diatonos*. For a recent discussion of Greek instruments and the difficulties in identifying them see W. D. Anderson, *Ethos and Educaton in Greek Music*, Index: *s.v.* aulos; barbiton; kithara. For an analytical bibliography of recent literature on Greek music, including musical instruments, see. R. P. Winnington-Ingram, "Ancient Greek Music 1932-1957," *Lustrum*, III (1958), 5-57, 259-60.

(Table 1—continued)

Hypate meson	Principalis mediarum	Highest of the tetrachord of the middle
Hypaton diatonos	Principalium extenta	Extended tone of the lowest tetrachord
Parhypate hypaton	Subprincipalis principalium	Next to the lowest of the lowest tetrachord
Hypate hypaton	Principalis principalium	Lowest of the lowest tetrachord
Proslambanomenos	Adquisitus	Added tone

Table 2. Consonances (*Symphoniae*)

Greek Name	*Latin Name*	*Translation*
Diatessaron	Ex quattuor	Fourth (lit., "from the four")
Diapente	Quinaria	Fifth
Diapason	Ex omnibus	Octave (lit., "from all")

The diatessaron consists of 4 notes; 3 steps; and $2^1/_2$ whole tones (*productio*; lit., "drawing out," lengthening"), or 5 semitones, or 10 quarter tones (*dieses*). This consonance is in the epitritic ratio (4:3). The diapente has 5 notes; 4 steps; and $3^1/_2$ whole tones, or 7 semitones, or 14 dieses. It is in the hemiolic ratio (3:2). The diapason has 8 notes; 7 steps; 6 whole tones, or 12 semitones, or 24 dieses. It arises from the diplasic ratio (2:1).[21]

There are fifteen octave species (*tropi*):[22] five principal ones, with a pair of secondary *tropi* attached to each of them. (Martianus' order is shown in Table 3.)

Table 3. Octave Species (*Tropi*)

Principal	Secondary
Lydian	Hypolydian
	Hyperlydian
Ionian	Hypoionian
	Hyperionian
Aeolian	Hypoaeolian
	Hyperaeolian
Phrygian	Hypophrygian
	Hyperphrygian
Dorian	Hypodorian
	Hyperdorian

[21] Cf. Nicomachus 2. 26; Theon, ed. Hiller, p. 56; Calcidius 94-96; Macrobius *Commentary* 2. 1. 14-25; Cassiodorus 2. 5. 7.

[22] See Mountford and Winnington-Ingram, p. 586.

A certain compatibility[23] exists between some *tropi*: between hypodorian and hypophrygian; between hypoionian and hypoaeolian; and between hypophrygian and hypolydian. The middle tone (mese) of a lower *tropus* becomes the proslambanomenos of a higher *tropus*. Each of these *tropi* comprises five tetrachords. A tetrachord is then defined as a set of four tones, arranged in order, of which the two extreme or bounding notes must be consonant.[24] (932-35)

Martianus here expresses the hope that his introductory remarks will make the ensuing discussion more intelligible. His treatment of music actually consists of two distinct sections (see above, pp. 53-54). Up to this point he has been relying upon some unknown source or sources. Begining at Section 936 his discussion is almost wholly derived from Book I of Aristides Quintilianus' *Peri mousikēs* (3d-4th cent. A.D.) or a Latin translation of it.[25]

Taken in its broadest sense, the discipline *harmonia* is found to have three divisions: subject matter (*hylikon*), practice (*apergastikon*), and exposition (*exangeltikon* or *hermeneutikon*). Each division has three subdivisions. *Hylikon* here refers to consonant elements that are continuous and similar: melody (*harmonica*), measure (*rhythmica*), and words (*metrica*).[26] *Apergastikon* is subdivided as follows: musical composition (*melopoeia*), choice of pitch (*lēpsis*), and relation of pitches (*plokē*).[27] The subdivisions of *exangeltikon* are instrumental music (*organikon*), vocal music (*ōdikon*), and recitation (*hypokritikon*).[28] Each of these will be explained later, in its appropriate place.[29] (936)

[23] Martianus' expression, *amica concordia*, is not a musical term.

[24] Greek musical theorists called the extreme notes *phthongoi hestōtes* [fixed notes], and the inner notes *phthongoi kinoumenoi* [movable notes]. See *ibid.*, p. 587.

[25] R. P. Winnington-Ingram, recent editor of Aristides Quintilianus *De musica* (Leipzig, 1963), p. xxii, does not agree with Deiters that Martianus had a text of Aristides before him.

[26] Cf. Aristides 6. 17.

[27] Cf. Aristides 28. 10 - 29. 8. Martianus' divisions differ from those of Aristides. On Aristides' divisions and definitions see D. B. Munro, *The Modes of Ancient Greek Music*, pp. 62-65.

[28] Cf. Aristides 6. 22.

[29] A promise only partially kept. Martianus does deal later with *melopoeia* (966), *lēpsis* (994), and *plokē* (958).

Voice production is divided into three classes: continuous, discrete, and intermediate. Continuous pitch variation is used in flowing conversation; discrete pitch variation in music; the intermediate movement of the voice is used in the recitation of poetry.[30] (937)

The subject of harmonics has seven headings or topics (938): tones (*sonus*); intervals (*spatium*); systems (*systema*); genera (*genus*); character of systems, i.e., keys (*tonus*); change of systems (*commutatio*); and melody construction (*modulatio* or *melopoeia*). (The remainder of Martianus' section on music is devoted to defining and explaining the seven terms.[31] He uses both the Latin and the Greek forms.)

A tone (*sonus; phthongos*) refers to a particular musical (*modulata*) production of the voice.[32] A tone of definite pitch is for Harmony what the point is for Geometry and the monad for Arithmetic. Tension (*intentio; tasis*) causes the voice to be produced and the tone to continue. The very word *phthongos* serves as an example of its definition, in the raising and lowering of the pitch of the voice as it pronounces the word. Tension and relaxation of the voice are active, high and low pitch passive. *Productio* is the movement of the voice from a lower to a higher pitch (*epitasis*) or the reverse (*anesis*). Deepness of tone is produced when the breath is drawn from deep within, sharpness of tone from the forepart of the mouth.[33] (938-40)

At this point Martianus returns to the subject of *tonoi* (941-44),[34] reminding his readers that he had enumerated them above. (Earlier [931][35] he had actually counted eighteen keys; now he lists twenty-eight.) The unknown authority on whom Martianus was relying

[30] Cf. Aristides 5. 25 - 6. 7. Regarding Aristides' introduction of a third or intermediate movement, Munro, pp. 116-17, conjectures that "the Greek language had in a great measure lost the original tonic accents, and with them the quasi-melodious character which they gave to prose utterance."

[31] Tones (939-47); intervals (948-53); systems (954); genera (955-59); character of systems, i.e., keys (960-63); change of systems (964); and melody construction (965-66). This is the standard classification of the subject, taken from Aristides 7. 8-12. The terms in Greek are: *phthongos, diastēma, systēma, genos, tonos, metabolē, melopoeia.* Cleonides opens his handbook *Introductio harmonica* with this classification and proceeds to define and elaborate on each of the topics.

[32] A translation of the definition of *phthongos* in Aristides 7. 15-16.

[33] Cf. Aristides 6. 27 - 7. 6.

[34] Cf. Aristides 7. 17 - 9. 12.

[35] See above, p. 207.

earlier counted two movable notes in each tetrachord: parhypate and diatonos (or lichanos) in the lower tetrachords, trite and diatonos in the upper. Aristides Quintilianus, whom Martianus is now following, enumerates twenty-eight *tonoi*, indicating the possible ranges of motion of the movable notes in the three genera.[36] Two *tonoi* are thus added to each of the five tetrachords, accounting for the discrepancy in the two lists.[37] (To illustrate, let us take one case, that of the movable notes between the fixed notes of the hypaton and meson tetrachords; see Table 4.)

Table 4. Movable Notes for Conjunction between Hypaton and Meson Tetrachords

According to System of Eighteen Tonoi	*According to System of Twenty-Eight Tonoi*
Hypate hypaton	Hypate hypaton
Parhypate hypaton	Parhypate hypaton
	Hypaton enharmonios
	Hypaton chromatikē
Hypaton diatonos	Hypaton diatonos
Hypate meson	Hypate meson

Tones (*phthongoi*) are either "fixed" (*stantes; hestōtes*) or "movable" (*vagi; kinoumenoi*). The movable tones, in ascending order, are called *barypyknoi*, *mesopyknoi*, and *oxypyknoi*. *Barypyknoi* are tones that occupy the lower range of the *pyknon*;[38] *mesopyknoi* occupy the middle range; and *oxypyknoi* occupy the upper range.[39] The words *pyknoi* and *spissi* indicate the crowded conditions of these tones.[40] *Apyknoi* is the name given to notes that stand in the place of any of the *pyknoi*. The former are not bound by any genus or rule.

Fixed and standing tones are either *apyknoi* or *barypyknoi*.[41] These

[36] Aristides 7. 18 - 8. 2.

[37] For a fuller explanation, with musical notations, see Henderson, p. 344.

[38] Gr. *pyknon* = Lat. *spissum* [compressed], i.e., the part of the tetrachord in which the intervals are small.

[39] On the close study of these theoretical microtones made by a group called the "Harmonists" see Henderson, p. 342; and Mountford and Winnington-Ingram, 587-88.

[40] At 950 Martianus notes that *pyknoi* pertain to dieses.

[41] Just above, Martianus classified *barypyknoi* as movable tones.

have a form and species of "principal" (*hypate*) notes. Some authorities refer to them as "stationary" (*statarii*) because they are unable to undergo changes. Movable or "wandering" (*vagi*) tones have larger or smaller steps. Some of these are called "like that next to the lowest" (*parhypatoeides*)[42] because they lie immediately above the first note of the lowest tetrachord; others are called "like the finger next to the thumb" (*lichanoeides*) from the finger (*lichanos*) that is used in producing the tones.[43] (945-46)

Some of these tones produce concords, others are dissonant and jarring. Those that are consonant are called *symphonoi*, those that are dissonant *diaphonoi*. Those that have a different designation of tone but the same pitch are called *homophonoi*.[44]

There are still other classifications of tones: (1) according to tension (*intentio*), resulting in a difference between high and low pitch; (2) according to the distance between tones; (3) according to conjunction of systems, when a tone belongs to one or more systems; [(4) according to region or locus of voice series][45] and (5) according to ethos, high tones signifying one kind of ethos, low tones another.[46] (947)

The second heading or topic of the subject of harmonics[47] is the interval, or *diastema*. A *diastema* is the space that lies between a higher- and a lower-pitched tone.[48] Small intervals are found in the

[42] See above, p. 207, esp. n. 19.

[43] Cf. Aristides 9. 13-26.

[44] See H. S. Macran, ed., *The Harmonics of Aristoxenus*, p. 237: "The term ὁμόφωνοι is applied to notes which differ in function but coincide in pitch. Thus the dominant of the key of D and the subdominant of the key of E fall alike on A."

[45] There is a lacuna in the manuscripts of Martianus. Since he is translating Aristides here (10. 11-12), the missing item is supplied, in brackets, from Aristides' text: τετάρτη ἡ κατὰ τὸν τῆς φωνῆς τόπον, ὅταν ὁ μὲν μείζονος, ὁ δὲ ἐλάττονος ᾖ τόπον. *Locus* (Gr. *topos*) is defined by Theon, ed. Hiller, p. 52, lines 9-11.

[46] On Aristides Quintilianus and ethos see W. D. Anderson, p. 80; and on ethos and pitch see *ibid.*, Appendix A, pp. 183 ff.

[47] See above, p. 211, for the seven topics.

[48] The first half of Isidore's definition (3. 20. 5) is the same as Martianus' but he is confused about the term: *diastema est vocis spatium ex duobus vel pluribus sonis aptatum.* Fontaine, I, 429, thinks that Isidore is confusing *diastema* and *systema.*

enharmonic dieses; an example of a large interval is the double octave, the largest interval that is found in the modes. Intervals are subdivided into composite and incomposite, or asynthetic. Incomposite intervals are those that occur by step.[49] Composite intervals are made up of intervals differing from each other. Some intervals are rational, others irrational; rational intervals can be represented as proportional, having a numerical ratio; irrational intervals do not have a ratio. Some intervals are consonant, others dissonant. Some are enharmonic, others chromatic, still others diatonic. Some are even (those that may be divided into equal parts, as a tone into two semitones); others are odd or excessive (perissa)—those that are divided into three semitones. Some intervals are crowded (*spissa*), others are roomier (*rariora*). The crowded ones are held together by dieses, the roomier ones by tones. Among these, some are consonant, others dissonant, the number of dissonant intervals being very great. The consonant intervals in each mode are six in number: the fourth (*diatessaron*; *ex quattuor*); the fifth (*diapente; de quinque*); the octave (*diapason; ex omnibus*), a consonance throughout the tonal space; the octave and fifth, or twelfth (*diapason kai diapente; ex omnibus et ex quinque*); and the double octave (*dis diapason; bis ex omnibus*). (The various intervals, with their notes, steps, tones, semitones, quarter tones (dieses), and the ratios represented, are shown in Table 5.)

Table 5. Intervals

Interval	*Notes*	*Steps*	*Tones*	*Half Tones*	*Quarter Tones*	*Ratio*
Diatessaron	4	3	$2^1/_2$	5	10	Epitritus (4:3)
Diapente	5	4	$3^1/_2$	7	14	Hemiolius (3:2)
Diapason	8	7	6	12	24	Diplasios (2:1)
Diapason kai diatessaron	11	10	$8^1/_2$	17	34	Diplasiepidimoiros (8:3)
Diapason kai diapente	12	11	$9^1/_2$	19	38	Triplasios (3:1)
Dis diapason	15	14	12	24	48	Tetraplasios (4:1)

[49] "Incomposite" because, in a given scale, no note can occur between them. See Macran, ed., *Aristoxenus*, p. 237.

The whole tone is in epogdous (9:8) ratio. Everywhere in the harmonic genus we are obliged to admit the diesis, which is a quarter tone.[50] (948-53)

The third topic of harmonics is systems. A system is a multitude of tones arising from the several modes.[51] There are in all eight absolute and perfect octave systems. (These, with their upper and lower notes, are given in Table 6.[52])

Table 6. Eight Octave Species of the Greater Perfect System

Species	*Upper Note*	*Lower Note*
Hyperdorian	Mese	Proslambanomenos
Mixolydian	Paramese	Hypate hypaton
Lydian	Trite diezeugmenon	Parhypate hypaton
Phrygian	Diezeugmenon diatonos	Hypaton diatonos
Dorian	Nete diezeugmenon	Hypate meson
Hypolydian	Trite hyperbolaion	Parhypate meson
Hypophrygian	Hyperbolaion diatonos	Meson diatonos
Hypodorian	Nete hyperbolaion	Mese

The fourth topic of harmonics is genera. A genus is defined as the division of tetrachords in a certain way. There are three genera of movement (*genera modulandi*): enharmonic, chromatic, and diatonic. The enharmonic genus has the smallest steps and crowded notes (*pyknoi*); the diatonic is roomier and has whole tones; the chromatic consists of half tones. Just as whatever lies between black and white is called "color" (Gr. *chroma*), so the name "chromatic" is applied to the genus lying between the other two. (The descending order of intervals, by tones or fractions of a tone, is shown in Table 7. The

[50] Cf. Aristides 10. 16 - 11. 23; Nicomachus 2. 26; Theon, ed. Hiller, pp. 51-52; Vitruvius 5. 4. 7-8; Macrobius *Commentary* 2. 1. 14-20; Calcidius 92-94; Cassiodorus 2. 5. 7.

[51] On the lack of clarity of Martianus' definition see Fontaine, I, 429, fn. 5.

[52] Martianus is here referring to the eight octave species of the Greater Perfect System (*systema teleion meizon*). See Mountford and Winnington-Ingram, p. 586; Munro, pp. 35-40. The ethnic names in my Table 6 are not given by Martianus but are found in Mountford and Winnington-Ingram's table. Martianus is not following Aristides' presentation (13. 4 - 15. 20) here.

ascending order is reversed.) Martianus observes that in his day the diatonic scale was the one in greatest use.[53] (955-57)

Table 7. Descending Order of Intervals

Enharmonic genus	2	1/4	1/4
Chromatic genus	1 1/2	1/2	1/2
Diatonic genus	1	1	1/2

The progression of melody is accomplished sometimes by *agōgē*, sometimes by *plokē*. *Agōgē* refers to ascending consecutive notes; *plokē* is progression by skip. Progression from a low to a high pitch is called "straight" or "direct" (*eutheia; recta*); from a high pitch to a low pitch "bending backward" (*anakamptousa; revertens*); another progression, called *peripherēs* or *circumstans*, accommodates itself to either sequence.[54] (958)

Although there are countless ways of subdividing a tetrachord, there are only six familiar ones (see Table 8).

Martianus has little to say about *tonoi*, the next topic of harmonics and a complicated subject. Aristides Quintilianus, his authority on

Table 8. Divisions of Tetrachords*

Tetrachord	*Intervals*		
Enharmonic	2	1/4	1/4
Soft chromatic	1 5/6	1/3	1/3
Hemiolic chromatic	1 3/4	3/8	3/8
Tonic chromatic	1 1/2	1/2	1/2
Soft diatonic	1 1/4	3/4	1/2
Shrill diatonic	1	1	1/2

* Martianus (959) does not reproduce the intervals from Aristides 17. 11-12. See Mountford and Winnington-Ingram, p. 587.

[53] Cf. Aristides 15. 21 - 16. 10; Vitruvius 5. 4. 3. Gaudentius in the fourth century A.D. reported that the enharmonic and chromatic scales had become obsolete. See J. F. Mountford, "Greek Music and Its Relation to Modern Times," *Journal of Hellenic Studies*, XL (1920), 38.

[54] These are the three kinds of *agōgē*, according to Aristides 29. 8-13. He defines *eutheia* as "ascending by consecutive notes"; *anakamptousa* as "descending by consecutive notes"; and *peripherēs* as "ascending by conjunction and descending by disjunction, or the reverse." And see Macran, ed., *Aristoxenus*, p. 267.

musical *tonoi*, gives three definitions and applications of the word: (1) "tension," its literal meaning;[55] (2) "magnitude" or "extension of voice" (*megathos poion phōnēs*); and (3) "key" or "transposition scale," in the Aristoxenian sense of the thirteen *tonoi*.[56] Martianus is here appropriating Aristides' second definition when he says that *tonus* is a "magnitude of space" (*spatii magnitudo*) and explains that the term is used because the voice is "stretched" over a space from one note to another, from the mese to the paramese,[57] for example, or, in the Lydian genus, from a note whose symbol is an upright iota[58] to one whose symbol is a zeta or a supine pi. (960)

Each octave species comprises five tetrachords, as noted above.[59] The extreme or bounding notes of each must be consonant. A tetrachord is a coherent and faithful concord of four tones arranged in order. Martianus then gives the bounding notes of the first four tetrachords, those of the second being lost in a lacuna.[60] (961)

Next Martianus comes to a discussion of pentachords, a subject that is not treated by any extant classical author on music.[61] We would like to know his source. It is unlikely that these were a Roman invention. The pentachords, like the heptachords, are five in number. The first is called the "lowest of the principals" because it begins with the proslambanomenos and ends with the hypate meson. The next pentachord in ascending progression begins with the hypate diatonos and ends with the mese.[62] The third is the pentachord of the conjuncts, which goes from the meson diatonos to the nete synemmenon. The

[55] The noun *tonos* from the Greek *tenein* [to stretch].

[56] See *ibid.*, p. 586.

[57] Aristides (9. 2-4) uses the same example of a *tonos*.

[58] For a representation of these symbols in the Lydian scale see Macran, ed., *Aristoxenus*, pp. 50-51.

[59] See above, p. 210.

[60] Kopp, in his edition, p. 748, fills the lacuna from Meibom's text and wonders why no editor before Meibom—neither Grotius, Scaliger, nor Vulcanius—filled the lacuna.

[61] Pseudo-Plutarch (*De musica* 1141f) quotes a poem of the comic playwright Pherecrates (5th cent. B.C.) in which pentachords are mentioned.

[62] *Constat a principalium extenta in mediarum illam, quae in Lydio iota rectum habet.* Martianus' statement about the second pentachord is incorrect; the range as he gives it, from the extended tone of the principal to the extended tone of the middle tetrachord, would not produce a pentachord.

fourth extends from the mese to the nete diezeugmenon. The fifth pentachord begins with the diezeugmenon diatonos and ends with the nete hyperbolaion. According to Martianus some authorities maintain that pentachords can begin with semitones, as happens in the case of the diatonic scale. He would like to point out, however, that the proslambanomenos cannot be found in groups that do not start with a whole tone, that is, in the tetrachords (these always begin with semitones). The second step in a pentachord is a half tone. (961-63)

Now we come to the topic of "change of system" (*modulatio*). *Modulatio* is the shift of the voice to another key or tonal system. The shift is accomplished in any one of four ways: (1) by genus, for example, from enharmonic to either the chromatic or the diatonic scale; (2) by system, as, for example, from the Greater Perfect to the Lesser Perfect System, or from the conjuncts to the disjuncts; (3) by key, when the melody is transferred from the Lydian *tonos*, say, to the Phrygian or some other *tonos*; (4) by melodic movement, when we shift from one species of *modulatio* to another, or from a virile melody into a feminine one.[63] (964)

The seventh and last topic of harmonics is melody construction (*melopoeia*). Melopoeia is the effect of completed *modulatio*. *Melos* is the result of high and low tones. *Modulatio* is the expression of many tones. There are three styles of melopoeia: *hypatoeides*, also called "tragic," consisting of deeper tones; *mesoeides*, called ["dithyrambic," keeping equable tones in the middle range; and *neto eides*, or "nomic," consisting of tones in the higher range. Other styles are recognized as well: "erotic," including a separate type, "epithalamic"; "comic";][64] and encomiological. These styles are also referred to as *tropoi*. (965)

Melody construction may differ in genus (enharmonic, chromatic, diatonic); in style or *tropos* (hypatoid, mesoid, netoid); in mode (Dorian, Lydian, etc.) . . .[65] Martianus concludes his treatment of music by

[63] Cf. Aristides 22. 11-26.

[64] There is a lacuna in the Martianus manuscripts. The matter in brackets is supplied from Aristides' text (30. 5-7), of which this passage is a translation.

[65] There is another lacuna in the text here. The missing matter may be found in Aristides 29. 2 - 31. 2. Aristides divides, subdivides, and defines many terms. We do not know how much of this passage Martianus included. He refers later

advising anyone who undertakes musical composition to consider the system first and then to mingle and compose his notes accordingly.[66] (966) The preceding two-thirds of the music portion of Book IX was a digest of the first twelve chapters of Book I of Aristides Quintilianus' *Peri Mousikēs*, with occasional use of some other unknown source or sources. The section on rhythm and metrics to follow was digested in like manner from Chapters 13-19 of Book I.

Martianus opens this section with definitions of the Greek and Latin words for rhythm (*rhythmos* and *numerus*). His definition of *rhythmos* is a translation of Aristides', with a slight addition;[67] his definition of *numerus* was derived from some unknown Latin source.[68] *Rhythmos* is a grouping of times that are appreciable to the senses and are arranged in some orderly manner. *Numerus* is an orderly arrangement of different measures, subordinated to time and having regard for proportion in modulation (*modulatio*), that is, in the amplitude of sound or in the raising and lowering of the voice.[69] There is a difference between rhythm and that which has become rhythmic (*rhythmizomenon*). The latter is the implementation of rhythms (*materia numerorum*), whereas rhythm is considered to be the artificer of rhythmic movement or a species of it (*artifex aut species modulationis*).

In its broadest sense rhythm is divided into three categories: visual, as seen in bodily movements; auditory, in movement of the voice or sound (*modulatio*); tactual as when a doctor feels the pulse of a pa-

---

(994) to the three divisions of melopoeia (*lēpsis*, *chrēsis*, and *mixis*) and to the *tropoi* of musical ethos (*sustaltikos*, *diastaltikos*, and *mesē*) but the last two *tropoi* are also missing in a lacuna at 994. These six terms must have been defined in the lacuna at 966, as they are by Aristides 29. 2-7; 30. 12-15. On the *tropoi* of ethos see Munro, pp. 62-64; Henderson, p. 375.

[66] Cf. Aristides 28. 8 - 30. 15.

[67] Aristides 31. 8: σύστημα ἐκ χαόνων; Martianus (Dick 516. 8-9): *compositio quaedam* ex sensilibus *collata temporibus*.

[68] This is the opinion of Dick, expressed in the apparatus of his text.

[69] Martianus 967 (Dick 516. 10-13): *numerus est diversorum modorum ordinata conexio, tempori pro ratione modulationis inserviens, per id quod aut efferenda vox fuerit aut premenda*. Aristides 31. 9-10 gives the elements of *rhythmos* as *thesis* and *arsis* [lowering and raising (of the voice)], *psophos* [noise, sound] and *ēremia* [rest, quietude].

tient. The auditory and visual categories are the important ones in harmony. Rhythm in visual and auditory contexts is seen (1) in bodily motion; (2) in sounds or the movement of the voice, occurring in proper ratios; and (3) in the recitation of words. Rhythms in recitation are resolved into syllables; in movement of sound or voice, into thesis and arsis; in bodily movement, into figures and *schēmata* of gestures.[70] (967-69)

There are seven divisions of rhythm: (1) time or unit of duration (*tempus*); (2) subdivisions of time into "in rhythm" (*enrhythmon*), "not in rhythm" (*arrhythmon*), and "rhythmoid" (*rhythmoeides*); (3) feet; (4) kinds of feet; (5) rhythmic movement or progression (*agōgē rhythmica*); (6) "shift" or "transfer" (*metabolē*); and (7) rhythmical composition (*rhythmopoeia*).[71]

For Martianus, as for his source, Aristides, a time is an indivisible unit, comparable to the point of geometricians or the monad of arithmeticians. In words time is found in the syllable; in movement of sound or voice (*modulatio*) it is found in an indivisible sound or space; and in bodily movement, in a *schēma*, which, to Martianus, is the very beginning of a bodily motion. There are also times that are composite (*tempus compositum; synthetos chronos*),[72] divisible into two, three, or four portions. A time may be divided this far and no farther. In this respect a time is like a tone: just as a tone may be divided into four dieses, so a time may be divided into four portions.[73] (970-71)

Of times that are grouped in meters some are "enrhythmic," others "arrhythmic," still others "rhythmoid." Enrhythmic times observe a definite proportion in their arrangement, as in the case of those combined in a double or a hemiolic ratio. Times are arrhythmic which obey

[70] Martianus defines *schēma* below (971) as the very first moment of bodily movement in a pattern. Aristides (32. 6-7) is clearer and more precise about the rhythmic elements in bodily movement: there are "figures" or "patterns" (*schēmata*) and their extremities, which are called "points" (*sēmeia*). Martianus derived his definitions and categories from Aristides 31. 8 - 32. 7.

[71] Aristides (32. 8-10) recognizes five divisions. The plausible reason suggested by Dick (in the apparatus of his text) for Martianus' seven divisions is that Martianus wished to have these divisions conform with the seven divisions of music. In order to get seven, he subdivided Aristides' first two categories, time and feet.

[72] Aristides 32. 25.

[73] Cf. Aristides 32. 8-30.

no rule and are combined without a definite ratio. Rhythmoid times observe rhythm in some places and spurn it in others. Some times are designated as "rounded" (*strongyla; rotunda*), others "very full" (*periplea*). *Strongyla* times run along more precipitately or readily than a normal pace demands; *periplea* times unduly retard the rhythm and are suspended in a slow pronunciation. To recapitulate, there are simple (*simplex*) and composite (*multiplex*) times, the latter also being designated *podica* [belonging to the foot].[74] (972-73)

The foot is the first progression of rhythm, combining proportionate and related sounds (*progressio per legitimos et necessarios sonos iuncta*). It has two parts: thesis and arsis. There are seven aspects of foot: (1) according to size or magnitude (*magnitudo; megethos*), some being simple (e.g., the pyrrhic), others composite (e.g., paeons and epitrites); simple feet consist of times; composite feet are resolved into simple feet; [(2) according to genus (e.g., hemiolic or duplicate); (3) according to composition, some simple (two-time), others composite (e.g., twelve-time), the simple ones being resolved into times, the composite ones into simple feet; (4) according to ratio, that is, the rational or][75] irrational combination of thesis and arsis; (5) according to varieties of division, the various ways of dividing composite feet into different kinds of simple feet; (6) according to *schēmata* resulting from division; and (7) according to *antithesis* (*oppositio*), when, of two feet taken together, the preceding has the greater time, the following the lesser time, or vice versa.[76] (974-76)

There are three genera of rhythm, sometimes referred to as dactylic, iambic, and paeonic, sometimes as equal (*aequalia*), double (*duplica*), and hemiolic (*hemiolia*). Some authorities add the epitrite to these. An example of equal rhythm is found in a thesis of one time and an arsis of one time. The double rhythm has the ratio of 2:1, both in syllables and times. Hemiolic rhythm has the ratio of 3:2. The ratio of 4:3 is present in the epitritic foot. The dactylic foot has a thesis and an arsis with the same number of times. The iambic belongs to the

[74] Cf. Aristides 32. 30 - 3. 11.

[75] The bracketed matter is missing from the Martianus manuscripts and is supplied from Aristides' text (33. 15-21), of which Martianus' text is a translation.

[76] The text of 976 is corrupt, and emendations may be made by comparing it with Aristides 33. 12-28, of which Martianus' text is a translation.

duplex genus, whether the ratio of arsis to thesis is 1:2, 2:4, or any other double ratio. The hemiolic or paeonic genus exhibits the ratio of 2:3, and the epitrite the ratio of 3:4. The equal genus begins with a diseme, containing 2 time units, and goes up to 16 times. The diseme is the first to comprise an arsis and a thesis. An example is the word leo.[77] The double genus begins with a triseme, containing 3 times, and extends to 18 times. The hemiolic begins with a pentaseme (5 times) and extends to 25. The epitritic begins with a heptaseme (7 times) and goes to 14. Its use is complicated.[78] (977-78)

Some rhythms are composite, others incomposite, still others mixed. Incomposite rhythms consist of only one foot, for example, the tetraseme. Composite rhythms consist of two or more feet. Others are mixed, being sometimes resolved into times, sometimes into feet, as in the case of the hexaseme. Of composite rhythms some are combined by syzygy, others by period. Syzygy is the coupling of two dissimilar feet. The combination of more than two dissimilar feet is by period.[79] (979)

The first genus of feet is the dactylic. In this genus six different incomposite feet occur (see Table 9).

Table 9. Dactylic Genus

Foot	*Composition*	*Notation*
Simple (lesser) proceleusmatic	Short syllable in thesis; short in arsis	∪ | ∪
Double (greater) proceleusmatic	Two shorts in thesis; two shorts in arsis	∪∪ | ∪∪
Greater anapaest	Long syllable in thesis; two shorts in arsis	— | ∪∪
Lesser anapaest	Two short syllables in arsis; long in thesis	∪∪ | —
Simple spondee	Long syllable in thesis; long in arsis	— | —
Greater (double) spondee	Two longs in thesis; two longs in arsis	— — | — —

[77] Dick thinks that the use of a Latin word as an illustration here indicates that Martianus was drawing upon a Latin source—a likely supposition, though Martianus could have been using his own example.

[78] Aristides (34. 14-15) says its use is rare. Cf. Aristides 33. 29 - 34. 15.

[79] Cf. Aristides 35. 3-17.

The lesser proceleusmatic is called "unintermittent" (*synechēs*)[80] because the quick succession of bounding syllables offers no opportunity for division. This foot should be used sparingly, for continuous use impinges upon the verse, which ought to be recited with some dignity. There are also two composite rhythms, produced by syzygy or copula, one called the Greater Ionic, the other the Lesser Ionic (see Table 10). The rhythms, incomposite and composite, belonging to the dactylic genus are thus seen to be eight in number.[81] (981-83)

Table 10. Composite Rhythms

Rhythm	*Composition*	*Notation*
Greater Ionic	Simple spondee; simple proceleusmatic	— — | ᴗᴗ
Lesser Ionic	Simple proceleusmatic; simple spondee	ᴗᴗ | — —

The dactyl gets its name from the observation that the arrangement of its syllables resembles a human finger;[82] the anapaest is so called because its order is reversed;[83] the proceleusmatic or pyrrhic is so called because of its repeated use in contests and children's games;[84] the spondee because of its frequent use at sacrifices.[85] The ionic gets its name from the unevenness of the measures, which have two long and two short syllables. Many listeners are restrained at hearing such meters.[86] (984)

The second genus of feet is the iambic. In this genus four simple, or incomposite, rhythms occur; two rhythms composite by syzygy; and twelve composite by period (see Table 11).

[80] Aristides (47. 7) calls it *aprepes* [unseemly] because it consists of a large number of short syllables.

[81] Cf. Aristides 34. 19 - 35. 2.

[82] Gr. *daktylos* [finger]; the dactyl, like the finger, has one long member and two short.

[83] Gr. *anapaistos* [struck back, rebounding].

[84] Gr. *prokeleusmatikos* [rousing to action beforehand].

[85] Gr. *spondē* [libation].

[86] Cf. Aristides 35. 18-26. Aristides says that the Ionic foot gets its name from its "vulgar associations," being used by the Ionians in their revels.

Table 11. Iambic Genus

Rhythm	*Composition*	*Notation*
	SIMPLE OR INCOMPOSITE	
Iamb	Short arsis; thesis double in length	ᴗ | —
Trochee	Doubled thesis; short arsis	— | ᴗ
Orthius	Four-time arsis; eight-time thesis	ᴗᴗ ᴗᴗ | ᴗᴗ ᴗᴗ ᴗᴗ ᴗᴗ
Semantic trochee	Eight-time thesis; four-time arsis	ᴗᴗ ᴗᴗ ᴗᴗ ᴗᴗ | ᴗᴗ ᴗᴗ
	COMPOSITE BY SYZYGY	
Trochaic bacchic (*bacchius ab trochaeo*)	Trochee first; iamb second	— ᴗ | ᴗ —
Iambic bacchic (*bacchius ab iambo*)	Iamb first; trochee second	ᴗ — | — ᴗ
	COMPOSITE BY PERIOD	
Comprising one iamb, three trochees		
Iambic trochee (*trochaeus ab iambo*)	Iamb in first position	ᴗ — — ᴗ — ᴗ — ᴗ
Bacchic trochee (*trochaeus a bacchio*)	Iamb in second position	— ᴗ ᴗ — — ᴗ — ᴗ
Trochaic bacchic (*bacchius a trochaeo*)	Iamb in third position	— ᴗ — ᴗ ᴗ — — ᴗ
Epitritic iamb (*epitritus iambus*)	Iamb in fourth position	— ᴗ — ᴗ — ᴗᴗ —
Comprising one trochee, three iambs		
Trochaic iamb (*iambus a trochaeo*)	Trochee in first position	— ᴗᴗ — ᴗ — ᴗ —
Bacchic iamb* (*iambus a bacchio*)	Trochee in second position	ᴗ — — ᴗᴗ — ᴗ —
Iambic bacchic (*bacchius ab iambo*)	Trochee in third position	ᴗ — ᴗ — — ᴗᴗ —
Epitritic trochee (*epitritus trochaeus*)	Trochee in fourth position	ᴗ — ᴗ — ᴗ — — ᴗ
Comprising two trochees, two iambs		
Simple trochaic bacchic (*simplex bacchius a trochaeo*)	Trochees first	— ᴗ — ᴗ ᴗ — ᴗ —
Simple iambic bacchic (*simplex bacchius ab iambo*)	Iambs first	ᴗ — ᴗ — — ᴗ — ᴗ
Middle trochee (*trochaeus medius*)	Trochees in middle	ᴗ — — ᴗ — ᴗ ᴗ —
Middle iamb (*iambus medius*)	Iambs in middle	— ᴗ ᴗ — ᴗ — — ᴗ

* Also called (986) "middle bacchic" (*bacchius medius*).

The iamb gets its name from the Greek verb *iambizein*, meaning "to detract" or "to assail."[87] The meter was used for lampooning, the name suggesting the poison[88] of malice or spite. The trochee gets its name from its swift turning, like that of a wheel.[89] The orthius is so called from the majestic[90] character of its thesis. The semantic trochee, being retarded in pace, gives indication of that drawn out and halting character.[91] The bacchics get their name from the fact that they most resemble the cries of bacchants and are most suited to songs of Bacchic revelry.[92] (985-88)

The third genus, the paeonic, comprises only two rhythms, both incomposite (see Table 12). The rhythms in this genus are not coupled by syzygy or period. The name diagyian is given to the first paeon because it has two separate members.[93] The second is called epibatic[94] because it consists of four members—two arses and two different theses.[95] (989)

There are several kinds of rhythms belonging to the mixed genera, produced by combining elements of the three genera above: two dochmiacs; three prosodiacs; two irrational choreics; and six other rhythms (see Table 13).

Table 12. Paeonic Genus

Rhythm	*Composition*	*Notation*
Diagyian paeon (*paeon diagyius*)	Long and short in thesis; long arsis	— ∪ | —
Epibatic paeon (*paeon epibatus*)	Long thesis; long arsis;* two longs in thesis; long arsis	— | — | — — | —

* Another lacuna occurs here, the missing words being supplied from Aristides 37. 7-9.

[87] Sentence copied almost verbatim by Isidore 1. 17. 4.

[88] Gr. *ios* [poison, venom]. This is Aristides' derivation (36. 26). The true origin of iamb is uncertain.

[89] Gr. *trochos* [wheel]. Cf. Isidore 1. 17. 3.

[90] Gr. *orthios* [straight, upright].

[91] Gr. *sēmantikos* [indicative of]: *semanticus sane quia cum sit tardior tempore, significationem ipsam productae et remanentis cessationis effingit.*

[92] Cf. Aristides 36. 1 - 37. 4.

[93] Gr. *diagyios* [two-limbed].

[94] Gr. *epibatos* [accessible, having a passage].

[95] Cf. Aristides 37. 5-12.

Table 13. Mixed Genera

Rhythm	*Composition*	*Notation*
	DOCHMIACS	
First dochmiac (*dochmiacus prior*)	Iamb; diagyian paeon	ᴗ – | – ᴗ –
Second dochmiac (*dochmiacus posterior*)	Iamb; dactyl; paeon	ᴗ – | – ᴗᴗ | – ᴗᴗᴗ
	PROSODIACS	
Three-feet	Pyrrhic; iamb; trochee	ᴗᴗ | ᴗ – | – ᴗ
Four-feet	Same as three-feet, with iamb added	ᴗᴗ | ᴗ – | – ᴗ | ᴗ –
		ᴗ – – ᴗ | – – ᴗᴗ
Of two syzygies	Bacchic; greater ionic	– ᴗᴗ – | – – ᴗᴗ
	IRRATIONAL CHOREICS	
Iamboid (*iamboeides*)	Long arsis; two theses	– | ᴗᴗ
Trochoid (*trochoeides*)	Two arses; long thesis	ᴗᴗ | –
	OTHER RHYTHMS	
Cretic (*creticus*)†	Trochee in thesis; trochee in arsis	– ᴗ | – ᴗ
Iambic dactyl	Iamb in thesis; iamb in arsis	ᴗ – | ᴗ –
Trochaic bacchic dactyl	Trochee in thesis; iamb in arsis	– ᴗ | ᴗ –
Iambic bacchic dactyl	Iamb in thesis; trochee in arsis	ᴗ – | – ᴗ
Iamboid choreic dactyl	Dactyl in thesis; dactyl in arsis	– ᴗᴗ | – ᴗᴗ
Trochoid choreic dactyl	Dactyl in thesis; anapaest in arsis	– ᴗᴗ | ᴗᴗ –

* These six rhythms are merely enumerated in the Martianus manuscripts (933). The analysis is lost in a lacuna and is supplied here from Aristides' text (38. 3-12).

† Not a true cretic.

The seventh and last division of rhythm[96] is *rhythmopoeia*, rhythmical composition. Rhythmopoeia is the composing of rhythms and the working out of all the figures to full perfection.[97] It has the same divisions as melopoeia:[98] *lēpsis* [perception], by which we understand

[96] See above, p. 220, for the seven divisions.

[97] Martianus 994: *condicio numeri componendi et omnium figurarum plena perfectio*; Aristides 40. 8: δύναμις ποιητικὴ ῥυθμοῦ, τελεία δέ ῥυθμοποιία ἐν ᾗ πάντα τὰ ῥυθμικὰ περιέχεται σχήματα.

[98] The divisions of melopoeia were missing in a lacuna above. See p. 218.

how much use to make of a certain rhythm; *chrēsis* [practice], according to which we arrange the theses and arses appropriately; *mixis* [mixing], by which we fit rhythms to each other artistically, if the occasion calls for it. There are also three styles or *tropoi* in rhythmopoeia, which, as we noted above,[99] is also the case with melopoeia: *systaltikos* [contracting]; *diastaltikos* [expanding]; and *hēsychastikos* [soothing].[100] Rhythm is masculine and melody feminine. Melody is an artistic form which, wholly lacking in postures and figures (*sine figura*),[101] is judged on its own. Rhythm, exercising manly activity, provides form, as well as various other effects, to sounds.[102] (994-95)

At this point Harmony brings her discourse to an abrupt close. The narrative setting, too, ends in a brief sentence. Harmony, humming a lullaby, joins Jupiter and the other gods as they accompany the bride and groom to the marriage chamber—"to the great delight of everyone." Two books were needed to describe the preparations for a wedding that ends here without a ceremony. This is no curtain of modesty. Martianus relished his few chances to indulge in lasciviousness. Once he had embarked upon the disciplines the setting became a mere device to draw his readers on. He concluded each of the discipline books with a hasty return to the narrative scene. The main purpose of his book now being accomplished, he is quick to drop the device.

The epilogue, an autobiographical poem twenty-seven lines in length, is the most interesting passage in the entire work. But for the flamboyant language and the resulting corruptions in existing manuscripts, we would know a great deal more than we do about the author's life.[103] Martianus closes the poem with a plea to his son to be indulgent as he reads the trifles of a doddering old man.

[99] The *tropoi* of melopoeia were also missing in the lacuna at 966. See *ibid.*

[100] The last two *tropoi* are missing in a lacuna here and are supplied from Aristides' text (40. 14-15). Previously Aristides had designated the third *tropos* of melopoeia as *mesē*; here it is *hēsychastikos.*

[101] Aristides (40. 22) describes melody as *aschēmatison* [wholly lacking *schēma* (referring to gestures and bodily movement)].

[102] Cf. Aristides 40. 8-25.

[103] For the autobiographical details of the poem, see above, pp. 9-20.

PART IV

Conclusion

Conclusion

THE TALE that Satire told Martianus on long winter nights, his eyes blinking and his aged head nodding as he struggled to keep awake, is now concluded. It is our melancholy task here to reflect upon the implications of the recitals of the quadrivium bridesmaids.

As we have seen, *The Marriage of Philology and Mercury* was one of the most widely circulated books of the Middle Ages. Its attractions for medieval readers are plain to see; but the book has had few readers in the modern world. We who examine it now are apt to lay it aside, as it has been laid aside in the past century, as a curious specimen of literary tastes and intellectual life at the close of the Empire. But *The Marriage* is not a curiosity, nor is it a museum piece to be examined in minute detail and in isolation. It is cast in the form of an allegory;[1] beyond the allegory contrived by Martianus, the work lends itself to a higher moral of greater significance for modern readers.

The long wintry nights and the senility of the author we may see as portending the Dark Ages. The marriage of the soaring and subtle Mercury, dear to classical poets and satirists, to a medieval personification of musty handbook learning represents the decay of intellectual life in the West during the later centuries of the Roman Empire. Mercury's full circle is closed: In his earliest role, before assuming the graces of the Hermes of Greek poetry, Mercury was a trickster god of Italian marketplaces. He is here once again reduced to to a bag of tricks–the rhetorical arts. Most of the characters in the book are females, and all the major ones are in fact medieval personifications. Philosophy, queen of the classical intellectual world, has become a "dignified woman with flowing hair" and "mother of scholars and of heroes great."[2] A few trite epithets suffice to describe her. The seven

[1] On allegory and moralization see Curtius, pp. 203-7.

[2] 131: *gravis crinitaque femina*; 576: *tot gymnasiorum ac tantorum heroum matrem.*

bridesmaids are decked with the stock accouterments and garments of medieval allegorical damsels, but the disciplines they offer as wedding presents are faint echoes from classical sources, from the *Timaeus* and the *De caelo*, from Cicero, Varro, and lesser men of learning. So much for the *dramatis personae*.

Now for the *moralisatio*. The interpretations to follow are not expressed here for the first time. The present study may be regarded as a case history reaffirming my earlier theses. Responses to those views have been generally consistent in their divergence: approbation from historians of science; almost complete silence from classical philologists. If the reticence of the latter stems from a feeling that I have denigrated Roman literature, a clarification is in order.

Historians of Roman literature have taken into their purview all Latin writings of any consequence in antiquity—an appropriate attitude to take with an early and fragmentary literature. Readers of their histories, however, are quick to appreciate the dichotomy between Latin belles-lettres and the nonliterature that bulks larger on our library shelves today. The belletristic masterpieces were created by men of extraordinary talent and have been savored and cherished by all subsequent generations to the present. Reading the works of compilers does not diminish our admiration and enthusiasm for master stylists like Vergil and Tacitus and artists like Propertius and Petronius. The men who had talent created poetry of exquisite beauty, and an impressive body of virile and incisive prose. Men devoid of talent, who aspired to literary careers, pillaged the writings of their predecessors and passed themselves off as men of great learning. These men deserve our censure. When respectable scientific subjects, together with the occult arts, were consigned by neglect to *viri doctissimi*, the doom of science in the West was sealed for a thousand years.

The turning of the way occurred in Republican Rome, and Cicero called the turn in Book I of his *De oratore*. He points out that the Greeks placed the philosopher and the specialist on the pedestals of their intellectual world, while the Romans more sensibly reserved the place of honor for the orator. The orator, he admits, cannot be expected to be an expert in every field of knowledge, though he is capable of mastering any subject through study, and his skill is of more consequence because he is more eloquent and better able to expound a subject than the "specialist or original thinker."[3] Cicero's ideal orator was not a master of Greek abstract and rigorously systematic disciplines; he prepared his briefs from derivative handbooks. His intellectual enthusiasms were for style and beauty in literature and rhetoric, not for science and philosophy, and the motivation for his professional researches lay in their applications to the arts of persuasion.

Cicero did not have the background or the temperament to transmit the specialized treatises of Hellenistic Greek writers. "Mathematics is obscure, abstruse, exact," he finds, "yet almost anyone who turns his mind to it can be a master in this field."[4] He himself had such an opportunity when he went off to his villa at Antium for an extended respite from politics in 59 B.C. He had been deceived, by a suggestion from his dear friend Atticus, into thinking that he could write a scientific treatise on geography. The project awaited his leisure. Atticus had sent him from Athens some Greek treatises by Eratosthenes and Serapion, and Cicero acceded. He applied himself to these books for a while but later admitted to Atticus in strict confidence that he could not comprehend a thousandth part of the matter.[5] Promises to complete the book followed, but nothing came of them: "geography is a dull subject and it gives a writer no opportunity to expound in a florid style."[6] On the other hand, if Varro, the most learned Roman of them all, had had the inclination to translate a dozen basic works of Euclid, Eratosthenes, Archimedes, and Hipparchus, he probably could have done so. He knew where to find Greek specialists to assist him, and manuscripts were available in abundance. Julius Caesar had the perceptiveness to appreciate the attainments of Alexandrian mathematicians and astronomers. His introduction of the Julian Calendar and composition of an astronomical treatise, regrettably lost, are evidence of his interests during a nine-month sojourn at Alexandria[7]—but his

[3] The very apology that Geometry made as she took over in Book VI (587) when Euclid and Archimedes were standing by and ready to present the subject.

[4] *De oratore* 1. 10.

[5] *Ad Atticum* 2. 4: *ex quo quidem ego, quod inter nos liceat dicere, millesimam partem vix intellego.*

[6] *Ibid.* 2. 6.

[7] For Moritz Cantor's appreciation of Caesar as a scientist see his *Die römischen Agrimensoren und ihre Stellung in der Geschichte der Feldmesskunst*, p. 78.

thoughts were more on Cleopatra and world empire. Moreover, Caesar was the one Roman author who used the austerely plain and objective style of the Attic masters of the classical period.

A society whose intellectual elite does not go beyond the level of books like Will Durant's *The Story of Philosophy* and Lancelot Hogben's *Science for the Citizen*, a society that breaks contact with original minds, as the Romans did, is doomed to intellectual decay. The way of the popular handbook, as it is digested and made more palatable for each succeeding generation, is inevitably downward. *The Marriage of Philology and Mercury* is a milestone in that downward course. Martianus stands almost at the halfway mark of Latin traditions. His success in epitomizing classical learning in the liberal arts and in transmitting it to the Middle Ages makes him our best index to the course of deterioration.

Had he lived in classical times, Martianus would not have been considered a *vir doctissimus*. His book, divested of its allegorical setting, is a school textbook. Its author, we conclude, must have spent a portion of his life as a schoolmaster. The liberal arts were basic in secondary education from Hellenistic times, the rhetorical trivium always attracting more attention than quadrivium studies.[8] Serious interest in scientific studies was limited to specialists in Hellenistic Greece as it was to *viri doctissimi* in Rome.

Martianus serves us best as an index if we make a frank evaluation of his work as a school textbook. To compare it with a product of our present-day schools is cruel but also enlightening; it is easy for us to point out defects and patronizing to remark upon the work's merits. Martianus' exposition of any of the four disciplines bears comparison with a term paper written by a high school senior of good standing who has a knack of turning out papers with a minimum of effort. Both are working under compulsions and handicaps. Both wish to cover the subject neatly, with the appearance of comprehending it. Martianus is handicapped by limited library resources and must deal with his subject in around eight thousand words. The high school scholar has probably given no thought to his paper during the football season and must write it in one week end, after a few hours of superficial

[8] See Marrou, *History*, pp. 183, 281-82; Bolgar, pp. 30-33.

research in the school library. There are two cardinal rules that both observe. The first is to avoid complexities at all times. The student has a Monday-morning deadline to meet. Martianus has not mastered his subject beyond the elements he presents; his excuse is that the subject is tedious and the reader disinterested. The second is to give the appearance of thoroughly mastering the subject. The student presents an imposing array of authoritative works in his bibliography, nearly all of them derived secondarily or only glanced at. He skillfully conceals his borrowings from his actual sources—encyclopedia articles or textbooks. Martianus never consulted the authorities he cites and never cites his actual sources, which were compendia and compilations of recent vintage.

All sorts of extenuations may be and have been offered in defense of writers like Martianus. The more common explanations point to the social and economic decay in the late Empire, the depredations of invading armies, and the Christian focus upon spiritual life, to the neglect of secular learning. These considerations account only for the degree of retrogression, however; they are not primary causes. If Martianus had been disposed to make a thorough search, he probably could have found a copy of Euclid's *Elements* or a minor astronomical work of Ptolemy's in a library at Carthage.[9] In choosing to depend upon recent popular sources, he was following the methods of Latin authors all the way back to the Republic.

Cultivated Romans of the last century of the Republic and the first century of the Empire were bilingual. Had they made the effort, they could have comprehended and translated the technical treatises of Hellenistic experts. For the more difficult sections they could have provided adequate paraphrases, as Calcidius did in the fourth or fifth century in his extended versions of Adrastus. The Romans in the heyday of the Empire traveled freely in the East and their government was in control of the intellectual and scientific centers at Athens, Alexandria, Rhodes, Pergamum, and Smyrna, with their vast library collections. They could have consulted scholars at the Greek centers, as Julius Caesar did at Alexandria. In fact, if there had been real in-

[9] See Walter Thieling, *Der Hellenismus in Kleinafrika*, pp. 27-30. On the intellectual life at Carthage at this time see B. H. Warmington, *The North African Provinces from Diocletian to the Vandal Conquest* (Cambridge, 1954), chap. VII.

terest in Greek technical treatises in Rome, the Greeks themselves would have prepared Latin versions of them. Instead Western Europe waited a thousand years, until scholars like Adelard of Bath, Gerard of Cremona, Michael Scot, and Herman of Carinthia produced Latin translations of the Greek masterworks, often from Arabic versions, and from texts corrupted and greatly depleted in intervening centuries. Even Varro, probably the best versed of ancient Romans in Greek scientific matters, did not reproduce the refinements of Greek science when he compiled the six hundred and twenty volumes credited to him. Every page of Martianus' quadrivium disciplines is a testimony to the dire consequences of these Republican attitudes.

Let us consider another example or two in passing. A basic tenet of classical Greek planetary theory was that Mercury and Venus had their orbits not about the earth but about the sun. That much of the heliocentric theory continued in vogue in the ancient Greek world after the abandonment of the complete system. Varro undoubtedly absorbed this doctrine from Posidonius' writings, but other classical Latin writers on astronomy allude to the theory without clearly comprehending it. Cicero, in his *De natura deorum* (2. 53), gives the maximum elongations for Mercury and Venus as one sign and two signs, stating that these planets sometimes precede and sometimes follow the sun; elsewhere he calls them the "sun's companions,"[10] but nowhere does he state that they go around the sun. Vitruvius, in a section on astronomy in Book IX, speaks of the two planets as "wreathed in their courses about the sun's rays as a center, making stations and retrogradations" (elsewhere "the powerful solar rays in a triangular position [trine aspect]" cause retrogradations of other planets[11]). Pliny refers to the stations of Mercury and Venus and gives their maximum elongations as 22° and 46°. His explanation of the stations of superior planets is a bizarre one. "When they are struck by a triangular ray of the sun they are stopped in their course and elevated straight upward, appearing to us to be stationary."[12] Macro-

[10] *Somnium Scipionis* 4. 2.

[11] *De architectura* 9. 1. 6, 12.

[12] *Natural History* 2. 38-39, 69-70. Yet Pliny cites as his authorities in astronomy the mathematicians Hipparchus, Sosigenes, Eudoxus, Eratosthenes, and Serapion, the last two being the writers who gave Cicero so much difficulty.

bius, in commenting on the passage in Cicero's *Somnium Scipionis*, gets involved in a lengthy dispute about the positions of Venus and Mercury with respect to the sun, attributing the cause of the confusion (not in his mind but surely in the reader's) to the higher and lower ranges of their orbits. Calcidius and Martianus, however, do understand and unequivocally describe the heliocentric motions of these planets, Calcidius because he is translating a Greek source here, and Martianus presumably because this doctrine had been preserved by his source from Varro's *Nine Disciplines*. Compare Calcidius' sophisticated discussion of the epicyclic motions of the planets with Pliny's nonsense, and the difference between Greek and Latin traditions becomes clear.

Or consider Isidore of Seville, the most learned man and most advanced scientific writer in Western Europe during his lifetime (*c.* 570-636). Let us quote a specimen of his science, taken from his bulky encyclopedia, the *Etymologies* (5. 31. 1-3):

> The word "night" [Lat. gen.: *noctis*] is derived from the word "to harm" [*nocere*] because it is harmful to the eyes. Night receives the light of the moon and stars to make it beautiful and to console men who work by night. It also provides a compensation for certain creatures that cannot bear the light of the sun. The alternating of day and night is designed for alternating sleep and wakefulness, night's rest offsetting the labors of day. Night comes about either because the sun is wearied by its long journey and when it reaches the edge of the heavens it becomes languid and its fires die out, or because it is driven beneath the earth by the same force that elevates its light above the earth and the resulting shadow of the earth causes night.

Who or what was responsible for the benighted character of Isidore's science? Was it the result of his absorption in spiritual matters? Certainly not, for his encyclopedia is filled with secular matters. Were the barbarian invasions responsible? Again no. The Vandals did not burn libraries. The libraries at Seville contained thousands of scrolls. If there were Greek works among them, Isidore would not have been able to read them. His familiarity with Greek was limited to words or phrases that came down in Latin context, to etymological glossaries, and to the tags of commentators. Isidore's encyclopedia is a compilation of badly deteriorated Latin traditions.

Science is hard to define and is conceived of on different levels. Scientific thinking of a rudimentary sort is involved in designing a fishhook. The book science we are dealing with in this study is also science on a low level. Book science is immediately responsive to upward and downward fluctuations of intellectual developments at large. On the other hand, the science of the fisherman and the science of original thinkers do not necesarily follow general intellectual trends. High-level science may reach a culmination long after other intellectual activities decline, as was the case in ancient Greece; and the fisherman's craft is not directly responsive to political, social, and intellectual developments. Neolithic man was in many respects an observing and rational creature and in deftness of hand probably exceeded modern man. There is no need for surprise if we discover that mining and glass-making techniques continued to improve during the barbarian invasions. Such technological developments are not a valid index of general intellectual progress. There is no disputing the statement that the level of science and secular philosophy in Western Europe during the first Christian millennium was low and that by the twelfth century it was distinctly higher. We are concerned at the moment with the reasons for the difference between the two levels.

It has been the fashion of twentieth-century scholarship to say that the term "Dark Ages" is outmoded. The author of the article with that title in the fourteenth edition of the *Encyclopaedia Britannica* says the term is no longer used. A recent, and perhaps the most vigorous, objector to the concept of a "Dark Ages" prefers to regard the Middle Ages as manifesting a shift of values, a change of moods of thought and expression, a cutting away of the dead wood of the decadent Empire.[13] Others have adduced radical improvements in horses' harnesses or in mining techniques, the beauties of stained-glass windows, or the revolutionary changes in town life as arguments refuting the concept of the "Dark Ages". Surely those who gave the term currency intended that it be applied to intellectual life as well as to town life or agrarian techniques. "Dark Ages" as applied to science and secular philosophy is not outmoded. Rather, the opinion of Charles Singer, an eminent historian of science who knew the early Middle Ages well, is the correct one. Addressing his remarks to those who

[13] William C. Bark, *Origins of the Medieval World*, pp. 2, 92, 98-107, 109, 200-2, 204.

were denying the validity of a Dark Age in science, he said: "For the painting of the Dark Ages no colors can be too dark."[14]

Trying to set dates for the beginning and the end of the Middle Ages is a baffling undertaking. Those who are satisfied with the comfortable notion that the Middle Ages began when the classical Roman Empire came to an end as a political entity can let the matter drop. But if, by Dark Ages, or Middle Ages, we mean retrogression and a distinctly lower level of scientific and philosophical thinking, the Middle Ages began in Western Europe in Italy during the Roman Republic. Once science and secular philosophy were transmitted to the Latin world in the first centuries B.C. and A.D., they became static, being virtually cut off from original Greek sources. They became the province of compilers and encyclopedists. Each generation digested, distilled, and garbled anew. Because writers preferred to use compilations prepared shortly before their own times, a deterioration in compiler literature, such as occurred in the successive geographical transmissions of Pliny, Solinus, Martianus, and Isidore, was inevitable. The trend of deterioration was generally continuous, though reversals occurred occasionally. A true Greek revival would have taken place in Italy during Theodoric's reign, early in the sixth century, when intellectual contacts with Byzantium were encouraged and reestablished—if Boethius had lived a long life and if had had several colleagues of similar competence and bent. He was undertaking to make the body of original Greek science and philosophy available to Latin readers through translations.[15] The Carolingian revival, on the other hand, was not a Greek revival but merely a renewal of interest in Latin compilations of an earlier period.[16] John Scot Eriugena was an exception, but

[14] "The Dark Age of Science," *The Realist*, II (1929), 283. T. E. Mommsen, in "Petrarch's Conception of the 'Dark Ages'," reprinted in his *Medieval and Renaissance Studies*, ed. E. F. Rice, Jr., (Ithaca, N.Y., 1959), pp. 106-29, traces concepts of the Dark Ages from Petrarch to the present.

[15] Charles Singer, *From Magic to Science*, p. 68, calls Boethius' failure to translate Aristotelian works on natural science a "world-misfortune" and names the three works which, in his opinion, would have changed the entire course of intellectual history: the *Historia animalium* and *De generatione animalium* of Aristotle, and Theophrastus' *Historia plantarum*.

[16] Remigius' commentary on Martianus, recently edited for the first time by Cora Lutz, reveals that Remigius at times had a better grasp of matters of Roman

his translations from the Greek did not markedly alter the character of the Carolingian revival.

Dates of inportant military or political events do not serve as demarcations in the course of these intellectual developments. Although Rome fell to Alaric in 410 and to Gaiseric in 455, studies and libraries were not appreciably affected.[17] Cassiodorus remarks, at the opening of his *Institutions* (*c.* 560), that the schools swarm with students pursuing secular learning. Boethius, in Rome a century after Alaric, engaged in intellectual pursuits that entitle him to rank with Cicero among the great Roman intellectuals. By contrast, Cassiodorus, Boethius' colleague in politics and intellectual matters, had both feet deep in the Middle Ages. Boethius was reading original Greek science and philosophy. Cassiodorus' *Institutiones* bears the indelible stamp of the Dark Ages.

The same considerations apply to attempts to set a time for the end of the Middle Ages. By the end of the twelfth century a true Greek revival had already occurred. Works of Hippocrates, Euclid, Aristotle, Archimedes, Apollonius of Perga, Heron of Alexandria, Galen, and Ptolemy had been translated from the Greek or from Arabic versions. The twelfth century saw the peak of the flood tide of translations that had begun two centuries earlier.

During the Middle Ages and the Renaissance, Latin was the language of men of learning, and it was hard for learned men to lose their respect for Latin authors. The mere fact that a Latin writer was an "ancient" entitled him to veneration. Even in the late Renaissance keenly perceptive scholars and scientists like Copernicus did not discern clearly between the original character of the Greek mind and

---

science than Martianus, whom he was sometimes able to correct or likely to clarify. A new question is posed: how was this knowledge transmitted to scholars in the Carolingian Age? Was it reposited in monasteries in Ireland, coming from entrepôts like Marseilles? The manuscripts needed to trace that transmission are almost totally lacking.

[17] When students nowadays use razor blades to excise entire chapters and articles from bound volumes and periodicals in college libraries, it is inappropriate to brand such acts as vandalism. The Vandals plundered a great deal but they evinced no such hostility toward public libraries.

the derivative character of the Latin writer. Thus the two bodies of literature intermingled congenially and there was no "battle of the books" on library shelves. The Latin masters found publication in incunabula editions as readily as did Latin translations of the Greek masters, and busts of Vitruvius and Celsus were likely to be placed alongside those of Archimedes and Ptolemy in frontispieces. Nevertheless, so far as intellectual history is concerned, the Middle Ages came to an end when the Latin tradition faded away, and the Renaissance began when the Greco-Arabic revival was initiated. These phenomena occurred gradually and overlapped each other during seven or more centuries.

So it was and always will be. The Greeks, the *Wunderkinder* of intellectual history, first propounded and pursued ideas for ideas' sake. The Roman penchant for doing rather than wondering represented a return to normalcy. And the conflict between the theoretical and practical approaches rages as hotly as ever today, among the policy-makers of education, government, industry, and the foundations.

The villain in our melancholy allegory was the popular handbook. These manuals did not originate in the Hellenistic Age. Like recipe books they had long been a part of daily life. The engineers of the pyramids undoubtedly used technical manuals not unlike recipe books. A foreman perched on the roof of a temple or a chef surrounded by pots, pans, and condiments did not want a theoretical treatise on the subject. Vitruvius' *De architectura* is an elaborate and diffuse treatment, designed for preparatory study and not for on-the-spot reference. But Vitruvius, being a Roman, could not escape the practices of compilers. The architects and surveyors of the Roman world used recipe books in the field.

A recent monograph by Manfred Fuhrmann points out that the Greeks, beginning with the Sophists and rhetoricians of the fifth century, developed the systematic teaching manual, or *Lehrbuch*, into a distinct genre. Fuhrmann is the first to have made a thorough and comprehensive study of a subject which, because of its unattractiveness, has been handled mostly by authors of encyclopedia articles or in dissertations dealing with individual works; doctoral candidates have found the handbooks ideally suited for dissertation studies. Fuhrmann's monograph is, however, too restricted in scope to be regarded as a definitive treatment of the subject.[18] By setting his *terminus ante quem at* A.D. 200 he missed an opportunity to introduce some of the best specimens of *Lehrbücher* for subjects that lack good examples in the classical period. He omits the quadrivium disciplines, except for Cleonides' manual on harmony; and Varro's *Nine Books of the Disciplines*, the key work in the Latin tradition, is treated in a single paragraph. Considering the importance of the handbook in intellectual history there is a need for a large work dealing with both the practical and the theoretical aspects.

The crowning irony of this tale lies in its implications for us today. The compiler was a transparent poseur. He was purportedly in touch with the ages. His most intricate and impressive revelations he got from the "Egyptians" and the "Chaldeans." These authorities are unassailable. Pliny says that the first one to understand the peculiar motions and behavior of the moon, including its eclipses, was Endymion.[19] In the Middle Ages Abraham, Moses, and Prometheus were cited as authorities on astronomy. The compiler makes sport of experts like Archimedes and Hipparchus. He brags that he will refute them and point out their fallacies.[20] He offers to be the first to settle a matter of higher mathematics. And this jackanapes has succeeded in imposing his frauds upon all generations of scholars to the present. His citations of authorities are still being taken seriously, and his learning is still regarded with undue respect.

This is not to say that the popular handbook is unworthy of our careful examination. In many fields the derivative compilation drove the original technical treatise out of circulation and is all that survives from which we can make a conjectural reconstruction, however insecure. Moreover, no matter how garbled a set of data or doctrines may have become in the hands of a compiler, there is always the remote possibility that they contain some bit of accurate information from an original work. Pliny, the master poseur of antiquity, is the

[18] The limitations, as well as the contributions of Fuhrmann's study, are discussed in my review essay on his monograph in *Latomus*, XXIII (1964), 311-21.

[19] *Natural History* 2. 42-43.

[20] It was a common practice of compilers to pretend to be disputing reputable authorities. It may be pointed out, to Martianus' credit, that he did not resort to their practice.

prime example of this observation. His *Natural History* still represents a vast collection of remarkably reliable data, as well as of curious and ridiculous nonsense. The task is to read him with discrimination.

The final implication is a perturbing one though too obvious to elicit a diatribe here. Latest advances in science, technology, and scholarly research, far from daunting the popularizer, seem to stimulate his imagination. And why should we, surrounded by bewildering innovations and specializations, be troubled with the refinements of new discoveries when a glib reporter can interpret the results or extract the kernel? The scientist's lament is that he finds time only for scanning abstracts; he is filled with qualms because the frontiers of his own field are receding beyond his ken.

But the situation in the humanities is an altogether different one; there the student drifts without an anchor. The many "vital" and "meaningful" experiences to which he is introduced do not include a single discipline with standards of accuracy. As he enters the college bookstore his glance is caught by flashy titles of study guides that reduce Dante and Aeschylus to handy précis, prepared by ready hands who themselves had little or no contact with the works in their original form.[21] If there comes a time when science manuals are written by humanities majors who know no basic science, a full circle of intellectual history will have been closed and we will be back again with the ancient Romans.

[21] This practice has gotten out of hand. Professors who once deplored the use of cribs and warned their students against them have themselves taken to writing study guides, reasoning that if students must use such books, they ought to be guided by the best authorities. For a recent report on the new profession see "Riding the Ponies," *Time*, 91 (Jan. 26, 1968), 74-75.

Appendix A

BIBLIOGRAPHICAL SURVEY OF THE SEVEN LIBERAL ARTS IN MEDIEVAL AND RENAISSANCE ICONOGRAPHY

MARTIANUS CAPELLA was the first author to allegorize the seven liberal arts. It was his descriptions of the maidens, together with the interpretations and embellishments introduced by ninth- to twelfth-century commentators on his work, that inspired illuminators, sculptors, and painters of the Middle Ages and the Renaissance. As E. F. Corpet pointed out in 1857,[1] with reference to Remigius, it matters not that the commentator was mistaken in his interpretations of Martianus' descriptions; what we really wish to know is the meaning that the Middle Ages attached to the attributes and to the symbols invented by the Middle Ages itself.

Though we lack evidence that manuscripts of *The Marriage* were illustrated from the beginning, scholars have recently conjectured that illustrations were present at an early date. Ludwig Heydenreich believes that the miniature representations of the liberal arts found in an early tenth-century manuscript (Paris, Bibl. Nat. MS. lat. 7900A) were derived from fifth- to seventh-century prototypes.[2] He compares certain features found in the Paris manuscript illustrations of Grammar, Dialectic, and Astronomy with those found in three other manuscripts of Martianus: Florence, Bibl. Med. Laur., San Marco 190 (11th cent.);[3]

[1] "Portraits des Arts Libéraux d'après les écrivains du Moyen Âge," *Annales archéologiques*, XVII (1857), 90.

[2] "Eine illustrierte Martianus Capella-Handschrift des Mittelalters und ihre Kopien im Zeitalter des Frühhumanismus," in *Kunstgeschichtliche Studien für Hans Kauffmann* (Berlin, 1956), pp. 60-61.

[3] The illustrations in this manuscript are discussed by C. Leonardi in "Illustrazioni e glosse in un codice di Marziano Capella," *Bullettino dell'Archivio paleografico italiano*, new ser., Vols. II-III, pt. 2 (1956-1957), pp. 39-60, esp. pp. 43-45.

Rome, Bibl. Vat., Urb. lat. 329 (15th cent.); and Venice, Bibl. Naz. Marc., lat. XIV. 35 (15th cent.). Rudolf Wittkower, in suggesting that thirteenth-century Italian miniatures of Solinus were derived from sixth- to seventh-century archetypes, remarks that Martianus too was probably illustrated at an early date.[4] Adolf Katzenellenbogen points to a miniature of the quadrivium in a Boethius manuscript written for Charles the Bald as the oldest extant representation of the liberal arts.[5] This miniature, he observes, embodies the literary tradition of Martianus and obviously followed earlier examples.

The first record of a pictorial representation of the seven arts, bearing resemblances to Martianus' depiction of them, is found in a poem entitled *De septem liberalibus artibus in quadam pictura depictis*,[6] composed by Theodulf, bishop of Orléans at the time of Charlemagne. There are several other medieval poems describing paintings and mosaics of the liberal arts.[7]

Paolo d'Ancona, Émile Mâle, and Raimond van Marle provide a general background and comprehensive survey of Martianus' influence upon medieval and Renaissance art, together with clear illustrations and plates of the examples discussed. D'Ancona's lengthy article[8] highlights the separate panels of the arts by Andrea Pisano and pupils on

[4] "Marvels of the East: A Study in the History of Monsters," *Journal of the Warburg and Courtauld Institutes*, V (1942), 171.

[5] "The Representation of the Seven Liberal Arts," in *Twelfth Century Europe and the Foundations of Modern Society*, ed. by Marshall Clagett, Gaines Post, and Robert Reynolds (Madison, Wis., 1961), p. 41.

[6] *Monumenta Germaniae Historica, Poetae Latini Aevi Carolini*, I (Berlin, 1891), 544-47; Migne, Vol. CV, col. 333. The poem is discussed by Leonardi, "Nuove voci poetiche tra Secolo IX e XI," *SM*, ser. 3, II (1961), 159-61; and Roger Hinks, *Carolingian Art* (London, 1935), pp. 151-52.

[7] See M.-T. d'Alverny, "La Sagesse et ses Sept Filles: Recherches sur les allégories de la Philosophie et des Arts Libéraux du IXe au XIIe siècle," in *Mélanges dédiés à la mémoire de Félix Grat*, I, 253-64; M. L. W. Laistner, *Thought and Letters in Western Europe A.D. 500 to 900*, 2d ed., pp. 213-14; Leonardi, "Nuove voci poetiche"; É. Mâle, *Religious Art in France, XIII Century*, pp. 79, 81, n. 4; C. E. Lutz, "Remigius' Ideas on the Origin of the Seven Liberal Arts," *Medievalia et humanistica*, X (1956), 33-34; E. R. Curtius, *European Literature and the Latin Middle Ages*, p. 39.

[8] "Le rappresentazioni allegoriche dell'arti liberali," *L'Arte*, V (1902), 137-55, 211-28, 269-89, 370-85.

the campanile of the Cathedral of Florence; the seven figures of the liberal arts on the tomb of Robert of Anjou in Naples; the frescoes of the seven arts in the Spanish Chapel of Santa Maria Novella, Florence; relief figures by Nicola Pisano on the pedestal of the pulpit of the Cathedral of Siena; those by Giovanni Pisano on the pedestal of the pulpit of the Cathedral of Pisa; and those by Nicola and Giovanni on the baptismal font, Piazza del Municipio, Perugia; the relief panels of the liberal arts in the Tempio Malatestiano, Rimini; the bronze figures by Antonio Pollaiuolo on the sepulcher of Sixtus IV in St. Peter's, Rome; the Attavante miniatures in the Codex San Marco at Venice; the fresco by Sandro Botticelli of Lorenzo Tornabuoni and the seven liberal arts, painted for the Villa Lemmi (now in the Louvre); and the figures of the arts by pupils of Pinturicchio in the Borgia apartments in the Vatican.

Mâle deals at length with the various attributes and symbolic representations of Martianus' bridesmaids and some of the innovations introduced by poets and artists of the Middle Ages.[9] He focuses particular attention upon the figures of the façade sculpture and in the rose windows of the cathedrals at Chartres, Laon, Auxerre, Sens, Rouen, and Freiburg. Karl Künstle[10] and Fritz Baumgarten[11] discuss in detail the figures of the seven arts of the porch of the Freiburg cathedral. Good illustrations of the figures on the Royal Portal of the Chartres cathedral are provided by Étienne Houvet,[12] and the figures are discussed by Adolf Katzenellenbogen.[13] Viollet-le-Duc's careful drawings of the figures on the west portal of the Laon cathedral preserve details no longer visible.[14]

In an extended chapter entitled "Les Sciences et les Arts" van Marle treats the representations of the seven liberal arts as a separate genre,

[9] *Religious Art in France*, pp. 76-90.

[10] *Ikonographie der christlichen Kunst*, I (Freiburg im Breisgau, 1928), pp. 145-56.

[11] "Die sieben freien Künste in der Vorhalle des Freiburger Münster," *Schau in's Land*, XXV (1898), 16-49.

[12] *Cathédrale de Chartres: Portail Occidental ou Royal XIIe siècle* (Paris, 1921), pls. 62, 64-67, 69, 71.

[13] *The Sculptural Programs of Chartres Cathedral* (Baltimore, 1959), pp. 16-21.

[14] Eugène Viollet-le-Duc, "Arts (Libéraux)," *Dictionnaire raisonné de l'architecture française du XIe au XVIe siècle*, II (Paris, 1875), 1-10.

declaring them to be the most popular mode of representing the arts and sciences in the Middle Ages and the Renaissance.[15] Van Marle's examples are mostly the same as those used by d'Ancona: the pulpit of the Pisa cathedral, the baptismal font at Perugia, the relief panels on the Campanile of the Florence cathedral, the frescoes in the Spanish Chapel of Santa Maria Novella in Florence, the bronze figures on the tomb of Robert of Anjou in Naples, the figures on the tomb of Sixtus IV in St. Peter's, the Tornabuoni fresco of Botticelli,[16] the Pinturicchio frescoes in the Vatican, and the Attavante miniatures in the Venice codex. Van Marle also discusses a miniature of a manuscript of the *Hortus deliciarum*, showing Philosophy surrounded by the seven liberal arts, and the tapestry in the Quedlinburg cathedral containing narrative scenes from *The Marriage of Philology and Mercury*. A full discussion of the Quedlinburg tapestry, with plate illustrations, is found in Betty Kurth's *Die Deutschen Bildteppiche des Mittelalters*.[17] Katzenellenbogen, in a chapter entitled "The Representation of the Seven Liberal Arts,"[18] pays particular attention to the figures on cathedral façades, declaring those of the Royal Portal at Chartres to be the earliest. He also discusses the *Hortus deliciarum* miniature and the figures on medieval candelabra at length.

Julius von Schlosser[19] and Josef Sauer[20] present bibliographical surveys of the seven liberal arts in medieval and Renaissance iconography. Donald Lemen Clark's concise survey, based on a careful examination of the specimens, emphasizes the particular differences in symbolic representation of the dress and paraphernalia of Martianus' bridesmaids

[15] Raimond van Marle, *Iconographie de l'art profane au moyen-âge et à la renaissance*, II (The Hague, 1932), 203-79.

[16] Van Marle also discusses the Tornabuoni fresco in *The Development of the Italian School of Painting*, XII (The Hague, 1931), 129-32.

[17] (Vienna, 1926), Vols. I, pp. 53-67; II, pls. 12-22.

[18] See n. 5 above.

[19] "Beiträge zur Kunstgeschichte aus den Schriftquellen des frühen Mittelalters," *Akademie der Wissenschaften, Vienna, Sitzungsberichte; phil.-historische Classe*, CXXIII (1891), 128-54: "Die Darstellungen der Encyklopädie, in besondere der sieben freien Künste."

[20] *Symbolik des Kirchengebäudes und seiner Ausstattung in der Auffassung des Mittelalters*, 2d ed. (Freiburg im Breisgau, 1924), pp. 433-36.

in later ages.[21] Jean Seznec traces the influence of Martianus Capella in medieval and Renaissance mythographical traditions.[22] Von Schlosser includes figures of the liberal arts among the precursors of figures in Raphael's Stanza della Segnatura.[23]

An ingenious article by Rudolf Wittkower traces the innovations in depicting Grammar from Martianus to Hogarth.[24] Klibansky, Panofsky, and Saxl bridge the gap between Martianus and Albrecht Dürer in the "typus Geometriae."[25] And Heydenreich's article on "Dialektik" in the *Reallexikon zur deutschen Kunstgeschichte*[26] contains a thoroughgoing treatment of this bridesmaid in art history. For the other bridesmaids readers of the *Reallexikon* are referred to a forthcoming article on "Künste." Karl-August Wirth, who is now editing the *Reallexikon* with Ludwig Heydenreich, is engaged in preparing a study of the liberal arts in the art history of the later Middle Ages.[27] Michael Evans, at the University of London, is writing a doctoral dissertation with the provisional title "The Representation of the Artes in the Middle Ages, and Their Sources." It will treat of allegories of the liberal arts in literature and the visual arts from Martianus' time until about 1450.

[21] "The Iconography of the Seven Liberal Arts," *Stained Glass*, XXVIII (1933), 3-17.

[22] *The Survival of the Pagan Gods: The Mythological Tradition and Its Place in Renaissance Humanism and Art* (New York, 1953).

[23] "Giusto's Fresken in Padua und die Vorläufer der Stanza della Segnatura," *Jahrbuch der kunsthistorischen Sammlungen des allerhöchsten Kaiserhauses*, XVII (1896), 13-100.

[24] " 'Grammatica': from Martianus Capella to Hogarth," *Journal of the Warburg and Courtauld Institutes*, II (1938), 82-84.

[25] *Saturn and Melancholy*, pp. 308-45, pls. 101-8.

[26] Vol. III, pt. 2, cols. 1387-1400.

[27] I am grateful to Dr. Wirth for calling my attention to the Von Schlosser and Sauer articles, to the Katzenellenbogen chapter on the liberal arts, and to a lengthy study by A. Filangieri di Candida, "Martianus Capella et le rappresentazioni delle arte liberali," in *Flegrea*, II (1900), pp. 114-30; 213-29. I was unable to examine the last article.

Appendix B

HAPAX LEGOMENA AND *RARIORA* FROM BOOKS VI-IX OF *DE NUPTIIS PHILOLOGIAE ET MERCURII*

THE FOLLOWING list is based upon Lewis and Short's *Latin Dictionary* (hereafter cited as L-S), Souter's *Glossary of Later Latin to 600 A.D.*, and the *Thesaurus Linguae Latinae* (*TLL*). *Rariora* have been included in this list because of their possible interest to philologists. The names of other authors found using these words are placed in parentheses; readers will note the high incidence of North African writers in this category.

Key to Symbols Used

* Not found in either L-S or Souter
** Found in Souter; not in L-S
|| Not cited in *TLL* for Martianus
† Not in L-S with Martianus' meaning
\# Cited by L-S for another author, not Martianus
\## Cited by Souter for another author, not Martianus

**acronycho 880 (pseudo-Censorinus; Calcidius)
adoperte 894 (*TLL* reads: adoperta)
**agalmata 567
allubescat 726 (Plautus; Apuleius; pseudo-Ambrose)
**†ambitorem 814
antistitiam 893 (only *TLL*: antistitium)
apocatasticus 735
apodictica 706, 715 (Gellius)
*aquilonalia 838 (Vitruvius)
**arrythmon 970 (Atilius Fortunatianus)
assecutor 905 (Fulgentius)
astrificante 585
astrifico 584 (adj.)
astriloqua 808
astrisonum 911
astructio 724 (Caelius Aurelianus; Rufus; Claudianus Mamertus)
astruere 592, 814[1]

[1] Lewis and Short are mistaken in stating that this word is not used in a figurative sense before Bede. Both Martianus and Macrobius (*Commentary* 2. 14. 6) use the word figuratively. See *TLL* for other instances.

aulicae 905 (Ambrose)
autumnascit (or -nescit) 605
blandifica 888
bombinator 999
bupaeda 908 (Varro)
**Caelulum 838
cerritulum 806
||chordacista 924
circumvolitabilis 584
coactibus 814[2]
coemesin 996
**colliniatae 729
colorabilis 942
compositiva 945
conspicabunda 803
contigue 909
conubialiter 576
#copulatus 731 (Arnobius)
corusciferi 808
*crinale spicum 903[3]
**culminatis 914
cuncticinae 905
Cyllenidae 899
*decuriatus 2, 728 (Livy)[4]
demulcatus 807
desorbentis 804 (Tertullian)
desudatio 577 (Scribonius Largus; Firmicus Maternus)
dicabulis 809
†diffusio 661 (Fulgentius; Cassius Felix)
dimerso 886
diplasia 934, 951 (Vitruvius; Favonius Eulogius; Fulgentius)
||diplasiepidimoiri 952
discludere 813[5]
discussius 891 (Jerome)
disemo 978 (Marius Victorinus)
disgrego 913 (Augustine; Boethius)
distermina 714
||**ditonon 957 (Boethius)
||diutule 803 (Gellius)
doctificus 567 (Priscian)
dulcinerves 917
**edissecatur 735
effigientiae 922
egersimon 911
||##**encomiologica 965 (Servius; Sacerdos)
enrythmon 970 (Varro; Censorinus)
||Eratine 905
exaratione 637 (Sidonius;[6] Porphyrion; Gregory I)
excusamentum 807
##**exsudatione 804 (Caelius Aurelianus; Cassius Felix)
extramundanus 910 (Jerome)
farcinat 998 (Cassiodorus)
fastuosa 579; 898 (Petronius; Martial)
fidicinat 929
fontigenarum 908 (Dracontius)
||glaucopis 571
||*helicoeides 868
heptas 738
hexas 738
**hexasemo 979 (Marius Victorinus)
**hiatimembrem 805
hircipedem 906
iambicinum 992
**Iastius 935 (Cassiodorus)
ideali 816
imbrificabat 584
inchoamentorum 576, 627, 935
infatigatus 582

[2] L-S mistaken: "Rare and only in abl. sing."

[3] Not cited with Martianus' meaning of *caelibaris hasta*, a small spear, the point of which was used to divide the locks of a bride's hair.

[4] Not cited with Martianus' meaning: "a decad of numbers."

[5] L-S says: "Already obsolete in the time of Macrobius."

[6] With a different meaning. For other late instances see *TLL*.

insopibilis 910
intercapedinatae 921 (Fulgentius; Caelius Aurelianus; Cassius Felix)
interrivata 627
interrivatione 661
interulos 888 (Apul.)
intervibrans 586
intimationis 897 (Calcidius)
irrisoria 809 (Augustine, Cassianus)
#juge 937 (Prudentius)
Latmiadeum 919
latrocinaliter 642
lepidulus 576; 726; 807
luxa (adj.) 914
Lymphaseum 569
marcidulis 727[7] (Fulgentius)
meacula 813
mediatenus 683; 840; 864; 933
melopoeia 938 (Fulgentius)
metaliter 859
†Midinus 577
monochronon 982
mutescentia 910 (Codex Theodosianus)
noctividus 571
nuptialiter 705
objectationis 924[8]
octas 740
octasemi 985
ornamen 587
ortivus 608 (Manilius; Apuleius)
Paedia 578
palmulari 805
particulatione 953
pentas 735
pentasemo 978
#peragratio 879 (Cicero)
perendinatio 897
perflagratus 576
*Pitho 906
**planontas 850
plaustrilucis 912
praeambulis 905; 996
praecluis 897; 906
praemetata 811
*praesis 573
#probamenti 732 (Codex Theodosianus)
*proferae 803
**rabulationis 577
rapiduli 804
recursio 911
reglutinatis 586 (Prudentius)
repensatrix 898
revibravit 810
rhythmoides 970
rhythmopoeia 970
##saltabunda 729 (Gellius)
scopa 812
sibilatrix 906
spinescere 704
spumigena 915
**submedia 961 (Boethius)
susurramen 726 (Apuleius)
susurratim 705
tellustres 729
tetraplasia 953
tetras 734 (Tertullian)
tetrasemi 979
**toniaea 959 (Boethius)
transaustrini 608
trias 733
trigarium (as a number) 733
triplasia 952
trisemo 978
Tritonida 893
vernicomae 570
vibrabundus 880
vibratum (noun) 887
vividas (verb) 912
*vomentis 647

[7] Cf. Apuleius *Metamorphoses* 3. 20: *luminibus marcidis.*
[8] With the meaning "presentation."

Selected Bibliography

Alverny, M.-T. d'. "La Sagesse et ses Sept Filles: Recherches sur les allégories de la Philosophie et des Arts Libéraux du IXe au XIIe siècle," in *Mélanges dédiés à la mémoire de Félix Grat*, I (Paris, 1946), 245-78.

Ancona, Paolo d'. "Le rappresentazioni allegoriche dell'arti liberali," *L'Arte*, V (1902), 137-55, 211-28, 269-89, 370-85.

Anderson, W. D. *Ethos and Education in Greek Music: The Evidence of Poetry and Philosophy*. Cambridge, Mass., 1966.

Aratus. *Phaenomena*. Edited by E. Maass. 2d ed. Berlin, 1893; reprinted, 1955.

—— *Commentariorum in Aratum reliquiae*. Edited by E. Maass. Berlin, 1898; reprinted, 1958.

Aristides Quintilianus. *De musica*. Edited by R. P. Winnington-Ingram. Leipzig, 1963.

—— *Von der Musik*. Introduction, German translation, and commentary by R. Schäfke. Berlin, 1937.

Aristoxenus. *Harmonics*. Edited with translation, notes, introduction, and index of words by Henry S. Macran. Oxford, 1902.

Auerbach, Erich. *Literary Language and Its Public in Late Latin Antiquity and in the Middle Ages*. Translated from the German by Ralph Manheim. New York, 1965.

Bark, William C. *Origins of the Medieval World*, New York, 1960.

Bayet, Jean. *Histoire politique et psychologique de la religion romaine*. Paris, 1957.

Beazley, C. R. *The Dawn of Modern Geography*. 3 vols. London, 1897-1906.

Bede, the Venerable. *Opera de temporibus*. Edited by Charles W. Jones. Cambridge, Mass., 1943.

Boethius, Anicius Manlius Torquatus Severinus. *De institutione arithmetica libri duo. De institutione musica libri quinque. Accedit Geometria quae fertur Boetii*. Edited by G. Friedlein. Leipzig, 1867.

Boissier, Gaston. *Étude sur la vie et les ouvrages de M. T. Varron*. Paris, 1861.

Bolgar, R. R. *The Classical Heritage and Its Beneficiaries*. Cambridge, 1954.

Bonner, Stanley F. *Roman Declamation*. Liverpool, 1949.

Böttger, C. "Über Martianus Capella und seine Satira: Nebst einigen kritischen Bemerkungen," *Neue Jahrbücher für Philologie und Paedagogik*, Suppl. Vol. XIII (1847), 590-622.

Bunbury, E. H. *A History of Ancient Geography*. 2 vols. 2d ed., London, 1883; reprinted, New York, 1959.

Burkert, Walter. *Weisheit und Wissenschaft*. Nuremberg, 1962.

Calcidius. *Timaeus a Calcidio translatus commentarioque instructus*. Edited by J. H. Waszink. London and Leiden, 1962.

The Cambridge History of Later Greek and Early Medieval Philosophy. Edited by A. H. Armstrong. Cambridge, 1967.

The Cambridge Medieval History. Edited by H. M. Gwatkin and others. 8 vols. New York, 1924-1936.

Cantor, Moritz. *Die römischen Agrimensoren und ihre Stellung in der Geschichte der Feldmesskunst: Eine historisch-mathematische Untersuchung*. Leipzig, 1875.

Capella, Martianus. *See* Martianus Capella.

Cappuyns, Maïeul. "Capella (Martianus)," *Dictionnaire d'histoire et de géographie ecclésiastiques* (Paris, 1949), Vol. XI, cols. 835-48.

—— *Jean Scot Érigène: Sa vie, son œuvre, sa pensée*. Louvain and Paris, 1933.

Cassiodorus. *Institutiones*. Edited by R. A. B. Mynors. Oxford, 1937.

—— *An Introduction to Divine and Human Readings*. Translated with an introduction and notes by Leslie Webber Jones. New York, 1946.

Cicero, Marcus Tullius. *De natura deorum*. Edited by A. S. Pease. 2 vols. Cambridge, Mass., 1955-1958.

Clagett, Marshall. *Greek Science in Antiquity*. New York, 1955; reprinted, 1966.

—— "King Alfred and the *Elements* of Euclid," *Isis*, XLV (1954), 269-77.

—— "The Medieval Latin Translations from the Arabic of the *Elements* of Euclid, with Special Emphasis on the Versions of Adelard of Bath," *Isis*, XLIV (1953), 16-42.

Clark, Donald Lemen. "The Iconography of the Seven Liberal Arts," *Stained Glass*, XXVIII (1933), 3-17.

Cleomedes. *De motu circulari corporum caelestium libri duo*. Edited by H. Ziegler. Leipzig, 1891.

Columba, G. M. "La questione soliniana," in *Ricerche storiche* (Palermo, 1935) I, 171-352.

Commentaria in Aristotelem Graeca. Volume IV. Edited by A. Busse. Berlin, 1887.

Copp, F. H. "The Doctrine of Music and Rhythm in Martianus Capella, *De nuptiis Philologiae et Mercurii*." Rendered into English with an introduction and notes. M. A. thesis. Cornell University, 1937.

Corpet, E. F. "Portraits des Arts Libéraux d'après les écrivains du Moyen Âge," *Annales archéologiques*, XVII (1857), 89-103.

Courcelle, Pierre. *Histoire littéraire des grandes invasions germaniques*. 3d ed. Paris, 1964.

—— *Les Lettres grecques en occident de Macrobe à Cassiodore*. Paris, 1943.

Crombie, Ian M. *An Examination of Plato's Doctrines*. 2 vols. New York, 1962.

Cumont, Franz. *After Life in Roman Paganism*. New Haven, Conn., 1923; reprinted, New York, 1959.

Curtius, Ernst Robert. *European Literature and the Latin Middle Ages*. Translated from the German by W. R. Trask. New York, 1953.

Dahlmann, H. "M. Terentius Varro," in Pauly-Wissowa, *Real-Encyclopädie der classischen Altertumswissenschaft*, Suppl. Vol. VI (Stuttgart, 1935), cols. 1255-59.

Deiters, Hermann. *Über das Verhältnis des Martianus Capella zu Aristides Quintilianus.* (Programm des königlichen Marien-Gymnasiums in Posen für das Schuljahr 1880/81, No. 131.) Posen, 1881.

Della Corte, Francesco. *Enciclopedisti latini.* Genoa, 1946.

—— *Varrone: Il terzo gran lume romano.* Genoa [1954].

Deonna, W. *Mercure et le scorpion.* (Collections Latomus, 37.) Brussels, 1959.

Dick, Adolf. *Commentationis philologicae de Martiano Capella emendando altera pars. St. Gall* [no date].

—— *De Martiano Capella emendando.* Bern, 1885.

—— *Die Wortformen bei Martianus Capella: Als Nachtrag zu Georges Lexikon der lateinischen Wortformen.* (Beilage zum Programm der St. Gallischen Kantonsschule.) St. Gall, 1901.

Dicuil. *Liber de mensura orbis terrae.* Edited by G. Parthey. Berlin, 1870.

Diehls, H., and W. Kranz, eds. *Fragmente der Vorsokratiker.* 9th edition. Berlin, 1960.

Dijksterhuis, E. J. *The Mechanization of the World Picture.* Translated from the Dutch by C. Dikshoorn. Oxford, 1961.

Dill, Samuel. *Roman Society in the Last Century of the Western Empire.* New York, 1899; reprinted, 1958.

Dolch, A. K. *Notker-Studien I-III.* Borna and Leipzig, 1951-1952.

Douglas, A. E. "Clausulae in the *Rhetorica ad Herennium* as Evidence of Its Date," *Classical Quarterly*, n.s., X (1960), 65-78.

Drabkin, I. E. "Posidonius and the Circumference of the Earth," *Isis*, XXXIV (1943), 509-12.

Dreyer, J. L. E. *History of the Planetary Systems from Thales to Kepler.* Cambridge, 1906; reprinted as *A History of Astronomy from Thales to Kepler*, New York, 1953.

—— "Medieval Astronomy," in *Toward Modern Science*, edited by R. M. Palter, Vol. I (New York, 1961), pp. 235-56.

Duckett, E. S. *The Gateway to the Middle Ages.* Ann Arbor, Mich., 1938; reprinted, 1961.

—— *Latin Writers of the Fifth Century.* New York, 1930.

Duff, J. W. *Roman Satire: Its Outlook on Social Life.* Berkeley, 1936.

Duhem, Pierre. *Le Système du monde.* 5 vols. Paris, 1913-1917; reprinted, 1954.

Dunchad. *Glossae in Martianum.* Edited by Cora E. Lutz. Lancaster, Pa., 1944.

Erhardt-Siebold, E. von, and R. von Erhardt. *The Astronomy of Johannes Scotus Erigena.* Baltimore, 1940.

—— *Cosmology in the Annotationes in Marcianum: More Light on Erigena's Astronomy.* Baltimore, 1940.

Eriugena, John Scot. *See* John Scot Eriugena.

Euclid. *Euclidis latine facti fragmenta Veronensia.* Edited by M. Geymonat. Milan, 1966.

—— *The Thirteen Books of Euclid's* Elements. Translated with introduction and commentary by Sir Thomas Heath. 3 vols. Cambridge, 1926; reprinted, New York, 1956.

Eyssenhardt, Franciscus. *Commentationis criticae de Marciano Capella particula.* Berlin, 1861.

Ferrero, Leonardo. *Storia del pitagorismo nel mondo romano.* Turin, 1955.

Fischer, Balduinus. *De Augustini disciplinarum libro qui est de dialectica.* Jena, 1912.

Fischer, Hans W. *Untersuchungen über die Quellen der Rhetorik des Martianus Capella.* Breslau, 1936.

Fontaine, Jacques. *Isidore de Séville et la culture classique dans l'Espagne wisigothique.* 2 vols. Paris, 1959.

Fuhrmann, Manfred. *Das systematische Lehrbuch: Ein Beitrag zur Geschichte der Wissenschaft in der Antike.* Göttingen, 1960.

Fulgentius, Fabius Planciades. *Opera.* Edited by R. Helm. Leipzig, 1898.

Geminus. *Elementa astronomiae.* Edited by K. Manitius. Leipzig, 1898.

Gerbert. *The Letters of Gerbert, with His Papal Privileges as Sylvester II.* Translated with an introduction by Harriet Pratt Lattin. New York, 1961.

—— *Opera mathematica.* Edited by N. M. Bubnov. Berlin, 1899.

Glover, T. R. *Life and Letters in the Fourth Century.* Cambridge, 1901.

Goldat, George D. *The Early Medieval Traditions of Euclid's* Elements. Ph. D. dissertation. University of Wisconsin, 1957. (Ann Arbor, Mich.: University Microfilms, Inc., Publ. No. 20, 236.)

Gudeman. "Grammatik," in Pauly-Wissowa, *Real-Encyclopädie der classischen Altertumswissenschaft,* Vol. VII, pt. 2 (Stuttgart, 1912; reprinted, 1958), cols. 1785-87.

Halm, Carl, ed. *Rhetores Latini minores.* Leipzig, 1863.

Haskins, C. H. *The Renaissance of the Twelfth Century.* Cambridge, Mass., 1927; reprinted, New York, 1957.

—— *Studies in the History of Mediaeval Science.* Cambridge, Mass., 1927; reprinted, New York, 1960.

Heath, Sir Thomas. *Aristarchus of Samos.* Oxford, 1913.

—— *A History of Greek Mathematics.* 2 vols. Oxford, 1921.

Heiberg, J. L. *Geschichte der Mathematik und Naturwissenschaften im Altertum,* Munich, 1925.

Henderson, Isobel. "Ancient Greek Music," in *The New Oxford History of Music,* Vol. I: *Ancient and Oriental Music* (London, 1957), pp. 336-403.

Heron of Alexandria. *Opera quae supersunt omnia.* Vol. IV. Edited by J. L. Heiberg. Leipzig, 1912.

Heydenreich, L. H. "Eine illustrierte Martianus Capella-Handschrift des Mittelalters und ihre Kopien im Zeitalter des Frühhumanismus," in *Kunstgeschichtliche Studien für Hans Kauffmann* (Berlin, 1956), pp. 59-66.

Hillgarth, J. N. "The Position of Isidorian Studies: A Critical Review of the

Literature since 1935," in *Isidoriana: Estudios sobre San Isidoro de Sevilla en el XIV centenario de su nacimiento* (Leon, 1961), I, 11-74.

Hinks, David A. G. *Martianus Capella on Rhetoric*. Unpublished Ph. D. dissertation. Trinity College, Cambridge, 1935.

Hinks, Roger. *Carolingian Art*. London, 1935; reprinted, Ann Arbor, Mich., 1962.

Hipparchus. *Geographical Fragments*. Edited with an introduction and commentary by D. R. Dicks. London, 1960.

Hisperica famina. Edited with a short introduction and index verborum by F. J. H. Jenkinson. Cambridge, 1908.

Honigmann, Ernst. *Die sieben Klimata und die ΠΟΛΕΙΣ ΕΠΙΣΗΜΟΙ*. Heidelberg, 1929.

Isidore of Seville. *Etymologiarum sive Originum libri XX*. Edited by W. M. Lindsay. Oxford, 1911.

—— *Traité de la Nature*. Edited by Jacques Fontaine. Bordeaux, 1960.

Jeauneau, Édouard. "Note sur l'École de Chartres," *Studi medievali*, ser. 3, V (1964), 821-65.

—— "Le *Prologus in Eptatheucon* de Thierry de Chartres," *Mediaeval Studies*, XVI (1954), 171-75.

John of Salisbury. *Metalogicon*. Edited by C. C. J. Webb. Oxford, 1929.

John Scot Eriugena. *Annotationes in Marcianum*. Edited by Cora E. Lutz. Cambridge, Mass., 1939.

Jones, Charles W., ed. *Bedae Pseudepigrapha: Scientific Writings Falsely Attributed to Bede*. Ithaca, N. Y., 1939.

Jones, Percy. *The Glosses* De musica *of John Scottus Eriugena in the MS. Lat. 12960 of the Bibliothèque Nationale, Paris*. (Ph. D. dissertation, Pontifical Institute of Sacred Music, 1940), Rome, 1957.

Jürgensen, Johann. "De tertio Martiani Capellae libro," *Commentationes philologae seminarii philologiae Lipsiensis*, 1874, pp. 57-96.

Kennedy, George. *The Art of Persuasion in Greece*. Princeton, N. J., 1963.

Ker, W. P. *The Dark Ages*. Edinburgh, 1904; reprinted, New York, 1958.

Klibansky, R. E. F., E. Panofsky, and F. Saxl. *Saturn and Melancholy*. New York, 1964.

Kneale, William and Martha. *The Development of Logic*. Oxford, 1962.

Knowles, David. *The Evolution of Medieval Thought*. London, 1962.

Koch, Josef, ed. *Artes Liberales von der antiken Bildung zur Wissenschaft des Mittelalters*. Leiden and Cologne, 1959.

Kristeller, Paul Oskar. *Renaissance Philosophy and the Mediaeval Tradition*. (Wimmer Lecture 15.) Latrobe, Pa., 1966.

—— *Renaissance Thought: The Classic, Scholastic, and Humanist Strains*. New York, 1961.

—— *Renaissance Thought II: Papers on Humanism and the Arts*. New York, 1965.

Kühnert, Friedmar. *Allgemeinbildung und Fachbildung in der Antike*. (Deutsche

Akademie der Wissenschaften, Schriften der Sektion für Altertumswissenschaften, 30.) Berlin, 1961.

Kurth, Betty. *Die deutschen Bildteppiche des Mittelalters.* 3 vols. Vienna, 1926.

Laffranque, Marie. *Poseidonios d'Apamée.* Paris, 1964.

Laistner, M. L. W. "Martianus Capella and His Ninth Century Commentators," *Bulletin of the John Rylands Library*, IX (1925), 130-38.

—— *Thought and Letters in Western Europe A.D. 500 to 900.* 2d ed., Ithaca, N.Y., 1957.

Langbein, Wilhelm. *De Martiano Capella grammatico.* Jena, 1914.

Lasserre, François. *The Birth of Mathematics in the Age of Plato.* Translated from the French by Helen Mortimer and others. New York, 1966.

Latham, R. E., ed. *Revised Medieval Latin Word-List from British and Irish Sources.* London, 1965.

Le Moine, Fanny J. *A Literary Re-evaluation of the De nuptiis Philologiae et Mercurii of Martianus Capella.* Unpublished Ph. D. dissertation. Bryn Mawr, Pa., 1968.

Leonardi, Claudio. "I codici di Marziano Capella," *Aevum,* XXXIII (1959), 433-89; XXXIV (1960), 1-99, 411-524. Also published as a separate volume, Milan, 1960.

—— "Illustrazioni e glosse in un codice di Marziano Capella," *Bullettino dell'Archivio paleografico italiano*, new ser., Vols. II- III, pt. 2 (1956-1957), pp. 39-60.

—— "Intorno al 'Liber de numeris' di Isidoro di Siviglia," *Bullettino dell'Istituto storico italiano per il medio evo e Archivio muratoriano*, LXVIII (1956), 203-31.

—— "Nota introduttiva per un'indagine sulla fortuna di Marziano Capella nel medioevo," *Bullettino dell'Istituto storico italiano per il medio evo e Archivio muratoriano*, LXVII (1955), 265-88.

—— "Nuove voci poetiche tra secolo IX e XI," *Studi medievali*, ser. 3, II (1961), 139-68.

—— "Raterio e Marziano Capella," *Italia medioevale e umanistica*, II (1959), 73-102.

Lewis, Charlton T., and Charles Short, eds. *A Latin Dictionary.* Oxford, 1879.

Lewis, Clive Staples. *The Allegory of Love: A Study in Medieval Tradition.* Oxford, 1936; reprinted, New York, 1958.

—— *The Discarded Image: An Introduction to Medieval and Renaissance Literature.* Cambridge, 1964.

Liebeschütz, Hans. "Zur Geschichte der Erklärung des Martianus Capella bei Eriugena," *Philologus*, CIV (1960), 127-37.

Lippman, Edward A. "The Place of Music in the System of the Liberal Arts," in *Aspects of Medieval and Renaissance Music: A Birthday Offering to Gustave Reese*, edited by Jan LaRue, Martin Bernstein, Hans Lunneberg, and Victor Yellin (New York, 1966), pp. 545-59.

Löfstedt, Einar. *Late Latin.* Translated from the Swedish by James Willis. Oslo, 1959.

Lüdecke, Fridericus. *De Marciani Capellae libro sexto.* Ph. D. dissertation. Göttingen, 1862.

Lutz, Cora E. "The Commentary of Remigius of Auxerre on Martianus Capella," *Mediaeval Studies*, XIX (1957), 137-56.

—— "Remigius' Ideas on the Classification of the Seven Liberal Arts," *Traditio*, XII (1956), 65-86.

—— "Remigius' Ideas on the Origin of the Seven Liberal Arts," *Medievalia et humanistica*, X (1956), 32-49.

McGuire, M. R. P. *Introduction to Mediaeval Latin Studies*. Washington, D.C., 1964.

Macrobius. *Macrobius* [opera]. Edited by James Willis. 2 vols. Leipzig, 1963.

—— *Commentary on the Dream of Scipio*. Translated with introduction and notes by W. H. Stahl. New York, 1952.

Mâle, Émile. *Religious Art in France, XIII Century: A Study in Mediaeval Iconography and Its Sources of Inspiration*. London and New York, 1913; reprinted (as *The Gothic Image: Religious Art in France of the Thirteenth Century*), New York, 1958.

Manilius. *Astronomica*. Edited by A. E. Housman. Cambridge, 1932.

—— *Astronomicon liber II*. Edited by H. W. Garrod. Oxford, 1911.

Manitius, Max. *Geschichte der lateinischen Literatur des Mittelalters*. 3 vols. Munich, 1911-1931.

Mariotti, Scaevola. "De quibusdam Macrobii et Martiani locis ad codicum lectionem restituendis," Reale scuola normale superiore di Pisa, *Annali*, ser. 2, IX (1940), 196-97.

Marle, Raimond van. *Iconographie de l'art profane au moyen-âge et à la renaissance*. Vol. II. The Hague, 1932.

Marrou, H.-I. *A History of Education in Antiquity*. Translated from the French by George Lamb. New York, 1956.

—— *Saint Augustin et la fin de la culture antique*. 4th ed. Paris, 1958.

Martianus (Min[n]e[i]us Felix) Capella. *Martianus Capella* [*De nuptiis Philologiae et Mercurii*]. Edited by Adolf Dick. Leipzig, 1925.

—— *Martianus Capella* [*De nuptiis Philologiae et Mercurii*]. Edited by Franciscus Eyssenhardt. Leipzig, 1866.

—— *De nuptiis Philologiae et Mercurii et de septem artibus liberalibus libri novem*. Edited by U. F. Kopp. Frankfurt-am-Main, 1836.

Martin, R. M. "Arts Libéraux (Sept)," *Dictionnaire d'histoire et de géographie ecclésiastiques*, Vol. IV (1930), cols. 827-43.

Mates, Benson. *Stoic Logic*. Berkeley, Calif., 1961.

May, Fr. *De sermone Martiani Capellae (ex libris I et II) quaestiones selectae*. Marburg, 1936.

Migne, J. P., ed. *Patrologiae cursus completus, series latina*. 221 vols. Paris, 1844-1864.

Miller, Konrad. *Die ältesten Weltkarten*. 6 vols. Stuttgart, 1895.

Monceaux, Paul. *Les Africains: Étude sur la littérature latine d'Afrique*. Paris, 1894.

Morelli, Camillus. "Quaestiones in Martianum Capellam," *Studi italiani di filologia classica*, XVII (1909), 231-64.

Mori, Assunto. "La misurazione eratostenica del grado ed altre notizie geografiche della 'Geometria' di Marciano Capella," *Rivista geografica italiana*, XVIII (1911), 177-91, 382-91, 584-600.

Mountford, J. F. "Greek Music and Its Relation to Modern Times," *Journal of Hellenic Studies*, XL (1920), 13-42.

—— and R. P. Winnington-Ingram. "Music," *The Oxford Classical Dictionary* (Oxford, 1949), pp. 584-91.

Munro, D. B. *The Modes of Ancient Greek Music*. Oxford, 1894.

Nettleship, Henry. *Lectures and Essays on Subjects Connected with Latin Literature and Scholarship*. Oxford, 1885.

Neugebauer, Otto. *The Exact Sciences in Antiquity*. 2d ed. Providence, R. I., 1957.

Nicomachus of Gerasa. *Introductionis arithmeticae libri II*. Edited by R. Hoche. Leipzig, 1866.

—— *Introduction to Arithmetic*. English translation by M. L. D'Ooge, with studies in Greek arithmetic by F. E. Robbins and L. C. Karpinski. New York, 1926; reprinted, Ann Arbor, Mich., 1938.

Norden, Eduard. "Die Stellung der Artes liberales im mittelalterlichen Bildungswesen," in *Die antike Kunstprosa von VI. Jahrhundert v. Chr. bis in die Zeit der Renaissance*, 2d ed. (Leipzig, 1923), II, 670-84.

Notker Labeo. *Notkers des Deutschen Werke*, Vol. II: *Marcianus Capella, De nuptiis Philologiae et Mercurii*. Edited, from the manuscripts, by E. H. Sehrt and T. Starck. Halle, 1935.

Nuchelmans, Gabriel. "Philologie et son mariage avec Mercure jusqu'à la fin du XIIe siècle," *Latomus*, XVI (1957), 84-107.

Ogilvy, J. D. A. *Books Known to the English, 597-1066*. Cambridge, Mass., 1967.

The Oxford Classical Dictionary. Edited by M. Cary and others. Oxford, 1949.

Parker, H. "The Seven Liberal Arts," *English Historical Review*, V (1890), 417-61.

Pauly, August Friedrich von. *Paulys Real-Encyclopädie der classischen Altertumswissenschaft*. Edited by Georg Wissowa, Wilhelm Kroll, and others. Stuttgart, 1894- .

Pire, Georges. *Stoicisme et pédagogie*. Paris, 1958.

Pliny the Elder. *Histoire Naturelle, Livres I, II*. Edited and translated by Jean Beaujeu. (Collection Budé.) 2 vols. Paris, 1950.

—— *Natural History*. Edited with an English translation by H. Rackham and others. (Loeb Classical Library.) 10 vols. London and Cambridge, Mass., 1942-1958.

Préaux, Jean. "Le Commentaire de Martin de Laon sur l'œuvre de Martianus Capella," *Latomus*, XII (1953), 437-59.

Price, Derek J. de Solla. *Science Since Babylon*. New Haven, Conn., 1962.

Quintilian, Marcus Fabius. *Institutiones Oratoriae Liber I*. Edited by F. H. Colson. Cambridge, 1924.

Raby, F. J. E. *A History of Christian Latin Poetry from the Beginnings to the Close of the Middle Ages*, 2d ed. Oxford, 1953.

—— *A History of Secular Latin Poetry in the Middle Ages*. 2 vols. 2d ed. Oxford, 1957.

Remigius of Auxerre. *Commentum in Martianum Capellam.* Edited by C. E. Lutz. 2 vols. Leiden, 1962-1965.

Rhetorica ad Herennium. Edited and translated by H. Caplan. Cambridge, Mass., 1954.

Ritschl, Friedrich. "De M. Terentii Varronis Disciplinarum libris commentarius" (originally published as *Quaestiones Varronianae* [Bonn, 1845]), in *Kleine philologische Schriften (Opuscula philologica)*, III (Leipzig, 1877), pp. 352-402.

Robbins, F. E. "Posidonius and the Sources of Pythagorean Arithmology," *Classical Philology*, XV (1920), 309-22.

—— "The Tradition of Greek Arithmology," *Classical Philology*, XVI (1921), 97-123.

Robinson, Richard. *Plato's Earlier Dialectic.* Oxford, 1953.

Roscher, W. H. *Die hippokratische Schrift von der Siebenzahl.* Paderborn, 1913.

Sandys, John E. *A History of Classical Scholarship.* 3 vols. Cambridge, 1903; reprinted, New York, 1958.

Sanford, Eva M. "The Use of Classical Latin Authors in the *Libri Manuales*," *Transactions of the American Philological Association*, LV (1924), 190-248.

Sarton, George. *Introduction to the History of Science.* 3 vols. Baltimore, 1927-1948.

Schanz, M., and C. Hosius. *Geschichte der römischen Literatur.* Vols. III; IV, pts. 1-2. Munich, 1914-1922.

Schiaparelli, G. V. *I precursori di Copernico nell'antichità.* (Pubblicazioni del Reale osservatorio di Brera, III.) Milan, 1873.

Schulte, Karl. *Das Verhältnis von Notkers De nuptiis Philologiae et Mercurii zum Kommentar des Remigius Autissiodorensis.* Ph. D. dissertation. Münster in Westfalen, 1911.

Servius. *Servii Grammatici qui feruntur in Vergilii carmina commentarii.* Edited by G. Thilo and H. Hagen. Leipzig, 1881-1887; reprinted, Hildesheim, 1961.

Silverstein, Theodore. "The Fabulous Cosmogony of Bernardus Silvestris," *Modern Philology*, XLVI (1948), 92-116.

Singer, Charles. "The Dark Age of Science," *The Realist*, II (1929), 280-95.

—— *From Magic to Science.* London, 1928; reprinted, New York, 1958.

De situ orbis libri duo [Anon.]. Edited by M. Manitius. Stuttgart, 1884.

Solinus, Gaius Iulius. *Collectanea rerum memorabilium.* Edited by Theodor Mommsen. 2d ed. Berlin, 1895; reprinted, 1958.

Souter, Alexander, ed. *A Glossary of Later Latin to 600 A.D.* Oxford, 1949.

Stachelscheid, A. "Bentleys Emendationen von Marcianus Capella," *Rheinisches Museum*, XXXVI (1881), 157-58.

Stahl, W. H. *Roman Science: Origins, Development, and Influence to the Later Middle Ages.* Madison, Wisc., 1962.

—— "The Systematic Handbook in Antiquity and the Early Middle Ages," *Latomus*, XXIII (1964), 311-21.

—— "To a Better Understanding of Martianus Capella," *Speculum*, XL (1965), 102-15.

Stange, F. O. *De re metrica Martiani Capellae*. Ph. D. dissertation. Leipzig, 1882.

Strecker, Karl. *Introduction to Medieval Latin*. English translation and revision by Robert B. Palmer. Berlin, 1957.

Sundermeyer, Albrecht. *De re metrica et rhythmica Martiani Capellae*. Ph. D. dissertation. Marburg, 1910.

Tannery, Paul. "Ad M. Capellae librum VII," *Revue de philologie*, XVI (1892), 136-39.

— *Sciences exactes au moyen âge*. Vol. V of *Mémoires scientifiques*. Edited by J. L. Heiberg. Paris, 1922.

Taylor, H. O. *The Mediaeval Mind*. 4th ed. 2 vols. Cambridge, Mass., 1951.

Teuffel, W. S., and L. Schwabe. *History of Roman Literature*. Translated from the fifth German edition by G. C. W. Warr. 2 vols. London, 1891-1892.

Theon of Smyrna. *Expositio rerum mathematicarum ad legendum Platonem utilium*. Edited by E. Hiller. Leipzig, 1878.

Thesaurus Linguae Latinae. 8 vols.- Leipzig, 1900-.

Thielung, Walter. *Der Hellenismus in Kleinafrika: Der griechische Kultureinfluss in den römischen Provinzen Nordwestafrikas*, Leipzig, 1911.

Thomson, J. O. *History of Ancient Geography*. Cambridge, 1948.

Thorndike, Lynn. *A History of Magic and Experimental Science*. 8 vols. New York, 1923-1958.

Thulin, Carl. "Die Götter des Martianus Capella und der Bronzeleber von Piacenza," *Religionsgeschichtliche Versuche und Vorarbeiten*, Vol. III, pt. 1. Giessen, 1906.

Turcan, Robert. "Martianus Capella et Jamblique," *Revue des études latines*, XXXVI (1958), 235-54.

Uhden, Richard. "Die Weltkarte des Martianus Capella," *Mnemosyne*, III (1936), 97-124.

Ullman, B. L. "Geometry in the Mediaeval Quadrivium," in *Studi di bibliografia e di storia in onore di Tammaro de Marinis*, Vol. IV (Verona, 1964), pp. 263-85.

Vogel, C. J. de. *Pythagoras and Early Pythagoreanism*. Assen, 1966.

Waerden, B. L. van der. *Science Awakening*. Translated from the Dutch by Arnold Dresden. Groningen, 1954.

Walbank, F. W. "The Geography of Polybius," *Classica et mediaevalia*, IX (1947), 155-82.

Weinhold, Hans. *Die Astronomie in der antiken Schule*. Ph. D. dissertation. Munich, 1912.

Weinstock, Stefan. "Martianus Capella and the Cosmic System of the Etruscans," *Journal of Roman Studies*, XXXVI (1946), 101-29.

Wessner, Paul. "Martianus Capella," in Pauly-Wissowa, *Real-Encyclopädie der classischen Altertumswissenschaft*, Vol. XIV (1930), cols. 2003-16.

Westphal, R. *Die Fragmente und die Lehrsätze der griechischen Rhythmiker*. Leipzig, 1861.

Willis, James A. *Martianus Capella and His Early Commentators.* Unpublished Ph. D. dissertation. University of London, 1952.

Winnington-Ingram, R. P. "Ancient Greek Music 1932-1957," *Lustrum*, III (1958), 5-57, 259-60.

Wirth, Karl-August. "Eine illustrierte Martianus-Capella-Handschrift aus dem 13. Jahrhundert," *Städel-Jahrbuch*, new ser., II (1969).

Wittkower, R. "Marvels of the East: A Study in the History of Monsters," *Journal of the Warburg and Courtauld Institutes*, V (1942), 159-97.

Wright, J. K. *Geographical Lore of the Time of the Crusades.* New York, 1925.

Youssouf Kamal. *Monumenta cartographica Africae et Aegypti.* Vol. III. [Place of publication not given], 1933.

Index